M000201740

THE HEROIC AGE of DIVING

AMERICA'S UNDERWATER PIONEERS AND THE GREAT WRECKS OF LAKE ERIE

JERRY KUNTZ

excelsior editions

State University of New York Press
Albany, New York

Published by State University of New York Press, Albany

© 2016 State University of New York

Excelsior Editions is an imprint of State University of New York Press

For information, contact State University of New York Press, Albany, NY
www.sunypress.edu

Production, Ryan Morris
Marketing, Kate R. Seburyamo

Library of Congress Cataloging-in-Publication Data

Kuntz, Jerry.
    The heroic age of diving : America's underwater pioneers and the great wrecks of Lake Erie / Jerry Kuntz. — Excelsior editions.
        pages cm
    Includes bibliographical references and index.
    ISBN 978-1-4384-5962-2 (pbk. : alk. paper)
    ISBN 978-1-4384-5963-9 (e-book)
    1. Deep diving—United States—History—19th century.   2. Deep diving—Erie, Lake—History—19th century.   3. Salvage—United States—History—19th century.
4. Salvage—Erie, Lake—History—19th century.
5. Shipwrecks—Erie, Lake—History—19th century.   I. Title.

VM977.K86 2016
627'.703—dc23
                                                                          2015010181

10 9 8 7 6 5 4 3 2 1

The Heroic Age of Diving

Contents

Principal Figures in *The Heroic Age of Diving*   ix

Acknowledgments   xi

PART ONE: PRELUDE—THE PIONEERS

Chapter One
Submarine Armor (1820s–1840)   3

Chapter Two
An Awful Calamity (1841–1844)   11

Chapter Three
End of the Taylors (1840s–1850)   23

Chapter Four
The Marine Engineers (1840s–1852)   33

PART TWO: THE HEROIC AGE OF DIVING

Chapter Five
The *City of Oswego* (July 1852)   57

Chapter Six
Without Armor and With Armor (July 1852)   67

Chapter Seven
Mr. Wells's Safe (August–October 1852)   75

Chapter Eight
The *Erie* Jinx (1853)   83

Gallery of photos follows page 90

Chapter Nine
Harrington and the Diving Boat (October 1853–Spring 1854)   91

Chapter Ten
Boston Bliss (1854–July 1855)   97

Chapter Eleven
Race to the *Atlantic* (August–December 1855)   105

Chapter Twelve
The Safe Recovered (1856)   119

PART THREE: THE AFTERMATH

Chapter Thirteen
The Moving Panorama (1857–1860)   129

Chapter Fourteen
War (1861–1865)   139

Chapter Fifteen
Ends (1866–1879)   149

Afterword: Envoi (1871–1891)   165

Notes   171

Bibliography   187

Index   191

Principal Figures in
The Heroic Age of Diving

Divers

John B. Green
Elliot P. Harrington
Martin Quigley
James A. Whipple
Charles B. Pratt
John Tope

Engineers and Entrepreneurs

William Hannis Taylor
George W. Taylor
John E. Gowen
Thomas F. Wells
Albert D. Bishop
Benjamin S. H. Maillefert
Henry B. Sears
James Eads
Lodner Phillips

Magnates

Henry Wells
Eber Ward

The Treasure Hunter

Daniel D. Chapin

The Wrecks

Steamer *Erie*
Steamer *G. P. Griffith*
Steamer *Atlantic*
Propeller *City of Oswego*
Steamer *Lexington*
HMS *Hussar*
San Pedro de Alcantera

Acknowledgments

Being neither a diver nor a naval historian, during the preparation of this text I frequently relied on the experience and expertise of others. Often, I would approach these people out of the blue, without introduction or credentials, but I never once came across anyone unwilling or hesitant to offer information, support, and suggestions. In no particular order I would like to thank the following for their assistance:

Mike Fletcher, Chuck Veit, James Delgado, Jack Messmer, Mike Babiski, Leslie Leaney, Tim Runyan, James Vorosmarti, Leon Lyons, Holly Izard, Peter Dick, Josh Thacker, Jaclyn Penny, Robert Delap, Tim Corvin, Barbara Kittle, Alice J. Murphy, Neil O'Brien, Lisa Hoff, Don Shomette, Bill Waldrop, Art Mattson, Ralph Gowen, Alfred Y. Gowen, James Tertius de Kay, Adam Lovell, Earl Verbeek, Alvin Oickle, Dayle Dooley, Mike Gray, Cynthia Van Ness, Jim Tremble, Patricia Harris, Justin White, S. White, Martin M. Quigley, Mike Spears, and Steve Cohen and Kathy Gunter.

Part One

PRELUDE—THE PIONEERS

Chapter One

≈

Submarine Armor (1820s–1840)

The founding father of American underwater exploration, William Hannis Taylor, began his career as a pirate. It was a label he desperately protested. It was also a term that investors in his later career probably used in a more figurative sense. However, the literal accusation came first: in December, 1828, Captain Daniel Turner of the United States Navy's sloop-of-war USS *Erie* was convinced that twenty-one-year-old Taylor was guilty of piracy. Taylor's ship, the schooner *Federal*, was seized by Captain Turner near St. Barts in the Caribbean and Taylor was taken into custody. At about the same time, naval vessels of other European powers also seized suspect ships in the Caribbean and accused them of the same high crime. As Taylor sat in confinement on the USS *Erie*, the punishment he faced was the same as that meted out to many of those other accused marauders—death by hanging.

Taylor's defense was that he had been acting as a privateer captain serving the government of the United Provinces of the Río de la Plata, now known as Argentina. Taylor carried letters of marque from that government authorizing him to seize ships belonging to—or carrying cargos owned by—the Provinces' declared enemy, the Empire of Brazil. Taylor's voyage that ended abruptly in St. Barts had begun in the port of Buenos Aires. He had left there hastily on the eve of ratification of a peace treaty between the Empire of Brazil and the United Provinces.

The treaty had already been negotiated and announced, and effectively meant an end to Taylor's legitimacy as a privateer. Leaving Buenos Aires hours before the formal ratification gave him the excuse of not having received notice to desist raiding. Communication of orders still depended on news traveling via ship, so this excuse left Taylor several weeks to continue to take prizes. His detractors suggested Taylor set sail with no intent to ever return to Argentina with any of those prizes, but instead had meant to head north with the objective of plundering as much as he could before returning home to New Bern, North Carolina.[1]

At that time, privateering was an accepted occupation for American mariners. It held several attractions: the romance of travel to distant lands, the glory of battle, the prospect of easy riches, and the opportunity to serve as a senior officer without the years of service usually required for promotion. Some privateers may have even pursued a Byronic idealism in fighting for fledgling democracies. While in service to the United Provinces, Taylor's commander was Commodore George DeKay, another young American with many connections to New York City's literary community. DeKay would go on to be hailed as a great humanitarian for his actions later in life. Taylor's later activities suggest that decidedly less noble motivations led him to become a privateer.

Fortunately for the accused pirate, Captain Turner of the USS *Erie* had made some glaring procedural missteps in seizing Taylor and the *Federal*. The island of St. Barts was at that time a colony of Sweden. When Turner found Taylor's ship *Federal* at St. Barts, Taylor had already unloaded some freight from the ship he had seized. Unloading the cargo, instead of delivering the entire ship and its contents back to the United Provinces, was a violation of customary international law. Captain Turner contacted the colonial officials at St. Barts and demanded that Taylor and his ship be turned over to him. The local officials did not comply, asking for proof of the crime. Turner suspected the local officials were in business together with the privateers, and were intentionally stalling. Pressed for time and conscious of his orders to seize privateers, Turner decided to act without the consent of the colonial council. He was not an inexperienced officer; in fact he had been a hero of the Battle of Lake Erie in 1813. On his own initiative, Turner decided to commandeer the *Federal* and seize Taylor. At one point the USS *Erie* was fired upon by the colonial St. Barts harbor defenses, but Turner wisely refused to return fire.

The Swedish government, as might be expected, protested Captain Turner's violation of their sovereignty. Turner's other miscalculation was trusting the assurances of his captive. While detained, Taylor argued that he needed to retrieve the letters of marque that would be crucial to his defense in any court or tribunal. After Turner granted his release to retrieve those papers, Taylor disappeared and never returned to custody. A couple of weeks later, Taylor materialized back at his home in North Carolina, where he immediately let it be known that he would be heading straight to Washington, D.C., to protest the actions of Captain Turner.[2] In Washington, a court of inquiry was convened, and no less an authority than President Andrew Jackson—mindful of the need for good relations with Sweden—concluded that Captain Turner had overstepped his authority. Taylor returned to North Carolina a man vindicated on principle, but one who had lost his ship, the *Federal*. On the plus side, he had narrowly escaped the hangman, and he had observed something while marauding in the Caribbean that he thought might prove more profitable than privateering. What he saw were pearl divers.[3]

≈

No one in the year 1828 would have called pearl diving a promising field of endeavor. The pearl industry in the Caribbean had been in a long decline, leaving an infamous legacy of human exploitation and ecological damage. During the heyday of the Spanish Empire in the New World in the sixteenth and seventeenth centuries, pearl-fishing operations flourished, and bushels of the oyster gems were delivered to Old World treasuries. The work of harvesting was done first by coerced native labor, and later, after that labor source was exhausted, by imported slaves from Africa and other colonies. It was dangerous work with a high mortality rate.

Soon after pearl harvesting was set up on an industrial scale, divers were forced to move into deeper depths to find untouched oyster beds. As depths increased, so did the dangers. Free-diving pearl gatherers risked burst eardrums, jellyfish stings, hypothermia, eye damage from prolonged exposure to salt water, and drowning from shallow- and deep-water blackout. The greatest risk—and source of countless horror tales—was of shark attacks.[4]

During the early 1830s, Taylor mulled over the physical dangers and limitations of free diving and considered the possibility of employing an assisted-breathing diving apparatus. The concept was thousands of years old, and had been realized to some degree by diving bells. However, diving bells required large ships and support personnel, and were less than practical for pearl harvesting. In the 1820s, English brothers Charles and John Deane invented a helmet into which air could be pumped; the device was initially developed for use by firefighters. By the early 1830s, the Deanes had adapted their design for use under water. Aided by another engineer, Augustus Siebe, they fitted the helmet to a waterproof body suit and added a valve to the exhaust outlet, which proved to be the most critical feature. By 1836, the Deane diving dress was being marketed to the public.

One year later, in 1837, Taylor published a booklet titled *A New and Alluring Source of Enterprise in the Treasures of the Sea, and the Means of Gathering Them.*[5] In it, Taylor made references to the Deane apparatus. Whether he had seen or tried to obtain a Deane suit is not known, but later that year, he submitted his own diving-suit design to the U.S. Patent Office. When one newspaper writer suggested that he had copied the English efforts, he wrote a prompt reply to refute that assertion: "the one [diving dress design] established by me in New York is of entirely different principle and construction, and has never been known until used by me in New York."[6] Taylor's patent drawing suggested a cylindrical helmet that was different from, but not as strong as, the Deane and Siebe rounded design. Aside from the helmet, the major difference was that in Taylor's version, the arms, legs, and entire torso were covered by protective hoops of plate metal; in the Deane and Siebe design, only the upper shoulders and neck were encased by metal. Taylor called his outfit "submarine armor," a name that persisted for decades, even after most of the metal sheet covering had been abandoned in favor of an outer layer of duck canvas. Taylor devised the protective armor plating for the purpose he had in mind: pearl harvesting in shark-infested waters.

Taylor first tested his submarine armor in August, 1837, in the waters of the Hudson River "a few miles above the city."[7] The exact location is not known, but the widening of the river at Haverstraw Bay and the Tappan Zee would have provided gentler currents. He not only publicized the test, but invited newsman James W. Hale of the Tontine News Room to try out the equipment. The account of this demonstration clearly stated that Taylor's aim was to prepare for a pearl-fishing venture.[8] Therefore the initial public displays of his patent-pending equipment were not intended to garner investors to set up manufacture of the diving apparatus; rather, they were meant to attract investors in an expedition to South America.

Since only a handful of people could witness a shipboard demonstration, two months later Taylor brought his diving dress to a large wooden vat set up in New York City's leading entertainment venue, Niblo's Garden. He descended into the water and stayed down for part of an hour, "much to the amusement of a large number of spectators."[9] After a few more demonstrations, Taylor's associates convinced him that it would be easier to attract investors to a local manufacturing venture than to a fishing expedition thousands of miles away.[10] Moreover, pearl harvesting in territorial waters required the cooperation of the host country, and offered no long-term stable income. And so, in January of 1838, a notice appeared in newspaper classified sections announcing the formation of the New York Sub-Marine Armour Company.[11]

It appears that one of William Hannis Taylor's associates, a man named George W. Taylor (not directly related to W. H. Taylor), took the lead in pursuing the manufacturing business.[12] George Taylor had signed as a witness to W. H. Taylor's patent application, but his background, other than claiming New Jersey as his birthplace, remains a mystery. George W. Taylor realized that selling submarine armor to wrecking companies—and investing in specific ad hoc salvage operations—offered better business prospects than pearl harvesting. Consequently, in June of 1838, the Taylors personally began cargo-recovery operations on America's two most recent, infamous shipwrecks, the sailing ships *Bristol* and *Mexico*, both of which foundered near Rockaway Beach, Long Island, with heavy loss of life. These two wrecks had shocked the American public in 1836–37, a fact that the Taylors must have realized would help advertise their recovery efforts. They were able to recover some metal bar freight—enough to cover their expenses and provide a small profit—but publicity was their ulterior motive.[13]

At about the same time, the Taylors added a new selling point to their marketing efforts: the ability to place and detonate underwater explosives. The brilliant British military engineer, General Sir Charles William Pasley, had already used divers to accomplish this feat. Wrecked ships stuck in sandy or muddy seabeds often could not be freed without blasting away the sediment that trapped them. Once again, it is not known whether the Taylors copied the idea from the English or developed it independently, but staged explosions soon became a staple of their submarine armor demonstrations. William H. Taylor's restless mind turned toward experimenting with the use of electric batteries to detonate the explosives as an alternative to unreliable waterproof fuse lines.

Later in 1838, the Taylors headed south to Florida in an effort to open a new market for their wares: the U.S. government. To promote the military value of their equipment, they offered to demonstrate the raising of a private launch that had been sunk in the St. Johns River during the Second Seminole War. They arrived first in St. Augustine and offered the public their standard diving and detonation show.[14] However,

William H. Taylor also showed off a small engine powered by galvanic cells—one of the first electric-battery motors ever invented.[15] Taylor's mind must have raced with the possibilities of that motor; after this Florida salvage operation, he left partner George W. Taylor in charge of the diving business. By the next year, William Hannis Taylor was exhibiting his electric motor in European engineering competitions, and vowed that he would only return to America "by lightning"—that is, in an electric-powered ship.[16] With William Taylor off the stage, George W. Taylor then inherited the role of America's foremost diving entrepreneur. The new "Captain Taylor" never publicly mentioned William H. Taylor again, and did not correct news reports that ascribed the invention of the Taylor submarine armor to him rather than W. H. Taylor.

From 1839 through 1841, George W. Taylor traveled up and down the Atlantic coast conducting more demonstrations and recruiting local dealers to sell the submarine armor. He quickly became the first person people thought to contact whenever there was a horrific shipwreck. Therefore, when the luxurious, state-of-the-art passenger steamer *Lexington* caught fire and sank in Long Island Sound in January, 1840, Taylor was called in to help recover bodies trapped in cabins and packages of money belonging to Harnden's Express.[17] Adolph Harnden, one of two brothers involved in the express business, perished on the *Lexington*. James W. Hale, the news agent who in 1837 had been one of the first persons to try Taylor's submarine armor, was an associate of the Harndens. It is likely that Hale urged the surviving brother, William F. Harnden, to bring Taylor in to explore the wreck of the *Lexington*.

Taylor reported that he found the wreck at a depth of 114 feet, which was far deeper than any of his previously recorded dives. He was able to recover a piece of the boat with metal attached, but otherwise was not able to make progress while operating at that depth.[18] A few years of experience informed Taylor that wrecks seldom took place in optimal diving locations for his apparatus: they were sunk too deep, or in powerful currents, or encased in sediment. What he needed was a showcase shipwreck in shallower, calmer waters.

Chapter Two

≈

An Awful Calamity (1841–1844)

Many American ships have borne the name *Erie*, starting with the sloop-of-war launched in the wake of the Battle of Lake Erie in 1813. This was the same USS *Erie* captained by Daniel Turner in 1828 that captured the privateer William H. Taylor and his ship, the *Federal*. However, the most magnificent and tragic ship to bear that name was the side-wheel steamer *Erie*, built for Charles Manning Reed by M. Creamer in Erie, Pennsylvania, in 1837. The *Erie* was designed to carry both passengers and cargo on the booming Lake Erie routes, which were already served by over three dozen steamers.

Although the first steam-powered ship plied Lake Erie's waters as early as 1818, it was not until the Erie Canal across New York was completed in 1825 that market forces caused a sharp rise in the number of steamship lines. The Erie Canal quickly became a primary route for settlers headed for the Ohio and Mississippi Valleys and for freight produced in the west being sent back east. Buffalo became the main point of departure for immigrants destined for Cleveland, Toledo, or Detroit and also further up the Great Lakes to ports such as Chicago, Milwaukee, and Duluth.

The *Erie*, measuring 176 feet long and 27 feet wide, presented a deceptively slender profile. Its gross tonnage was 497 tons, and it was capable of carrying more than 300 passengers and crew, with additional freight. General Charles M. Reed, the owner, commissioned a painting of the ship shortly after it went into service, which depicted the steamer churning smoothly through roiling waters. In the painting, the *Erie*'s white hull, cabin, and smokestacks are complemented by gold and green trim, with the walking beam located on the deck painted green and black. In the background, a sailing ship is left far behind in the distance. Though unsigned, this depiction of the *Erie* is such a magnificent example of maritime art that today it's in the National Gallery of Art in Washington, D.C.

On August 9, 1841, over 300 passengers boarded the *Erie* in Buffalo for the trip to Cleveland, Ohio, with scheduled stops along the New York and Pennsylvania shoreline. The *Erie* appeared at its pristine best; it had just been repainted from bow to stern. The team of painters was still aboard with their supplies, bound for the port of Erie, Pennsylvania, where they were scheduled to repaint a sister ship of the *Erie*. The passengers had arrayed themselves throughout the ship—in private staterooms, in the steerage compartment, in the spacious men's and women's deck cabins, or along the exposed deck promenade. Over 200 of the travelers were German, Swiss, and French immigrants, who had already survived the lengthy, perilous trip across the North Atlantic. Lake Erie, on this summer day, offered the prospect of gentle transit across a shallow inland sea, a state-of-the-art ship with comfortable accommodations and a veteran crew led by Captain Thomas Titus and wheelman Luther Fuller.

Only a couple of hours into the journey, near Silver Creek, New York, the *Erie* encountered winds that generated high waves, forcing nearly all the passengers inside. The motion might have caused some of the freight to shift, including glass demijohns full of paint and varnish belonging to the painting crew. Those containers, shielded by a thin wicker covering, had been placed on the main deck close to the ship's smokestack, near a grating over the engine boiler. Suddenly, a sharp explosion was heard throughout the ship, sounding as if it originated on the boiler deck. A cloud of black smoke enveloped the entire middle section of the ship. The cargo officer rushed to notify the captain, then shouted for a fire brigade. However, the flames were so high and widespread that many of the crew and those passengers fortunate enough to be on deck immediately began to abandon the ship.

Only a hundred life preservers were on board, all located in the women's cabin. The fire had spread so quickly that access to that cabin was impossible. In desperation, some passengers tore out wooden fixtures and threw them into the lake. The shore was distant, about seven or eight miles away. The crew rushed to lower the few lifeboats, but they were swamped by the panicked passengers. Captain Titus ordered wheelman Fuller to head the ship toward shore. He saw another ship in the distance, and shouted to the remaining passengers that help was on the way.

Ultimately, with no other escape, nearly all the passengers threw themselves into the water. Captain Titus found himself alone aboard within a few short minutes. He jumped into a nearby ketch that was still upright, but filled with water. He saw the waters around the burning vessel full of protruding heads, crying for help, but many soon slipped under the waves. The wheelman Fuller had stayed at his station, steering even as the engines lost steam, but to no avail—the entire ship was in flames six miles from shore. That was close enough for another ship, the *DeWitt Clinton*, to observe the *Erie* on fire from the dock at Dunkirk, New York. The *DeWitt Clinton* was able to reach the first survivors in the water in about an hour. Of the nearly 350 passengers and crew of the *Erie*, fewer than a hundred survived.

Powerless and abandoned, the *Erie* drifted as it burned completely down to the water line. Waves lapped over it, and it began to list. The machinery and structure below decks was spared from the worst of the flames. The *DeWitt Clinton* and another ship that had arrived on the scene, the *Lady*, attempted to tow the *Erie* to port. As they made fast, the *Erie* slipped under the waves and fell to the lake bottom, sixty-six feet below the surface. The bodies of the victims lay scattered on the lake bottom around the sunken hulk, but within a few days, decomposition brought many of those to the surface or washed them up on shore.

In the immediate wake of the *Erie* disaster, the nautical painting that had been proudly commissioned for her launching was reproduced as an etching in newspapers to accompany news of the "awful calamity," but now it was redrawn with plumes of smoke and flames erupting from amidships. Descriptions from witnesses and survivors of the horrible scene were published in newspapers throughout the United States and Canada. The loss of the *Erie* struck a chord in a young nation largely composed of recent immigrants; the fact that so many died with their names unknown and far from their home countries only added to the sense of tragedy. The heroism of the wheelman Luther Fuller, who was thought to have perished while maintaining his station, in coming years inspired a popular folk legend, celebrated in prose and verse (with Fuller's name changed to John Maynard):

> A sailor, whose heroic soul,
> That hour should yet reveal,
> By name John Maynard, eastern born,
> Stood calmly at the wheel.
> "Head her southeast!" the captain
> shouts
> Above the smothered roar,—
> "Head her south-east without delay!
> Make for the nearest shore!"
> —from *John Maynard* by Horatio Alger[1]

The assessment of responsibility for the wreck was more short-lived. Blame for the fire was immediately and plausibly placed on careless handling of the highly flammable painters' liquids. Captain Titus and his officers were held blameless; an inquest concluded that they had performed their duty capably as the ship sank. There was no announcement from owner Charles Reed that he planned to recover the wreck, or even auction it to salvagers. Knowledgeable mariners concluded that little of any value could remain, and that, with the ship laying at sixty-six feet of depth, salvage operations would be difficult, costly, and not worth the effort. Months after the calamity, there was no hint that the wreck of the *Erie* would, in the coming years, become the obsession of the most intrepid souls in the nation.

≈

In early 1842, George W. Taylor was the leader of America's only recognized company of divers. The previous year, he had followed the reports of the *Erie* tragedy along with the rest of America. The news must have increased his awareness that he had not yet tried to market the Taylor submarine armor in the Great Lakes area. After his mentor William H. Taylor went off to Europe in 1839, George W. Taylor worked with others to improve the apparatus. Foremost among these partners were two of the Goodyear brothers, Robert and Henry Goodyear, siblings of Charles Goodyear. All of the Goodyear brothers were in the India rubber business, but Charles immersed himself in developing vulcanized rubber while Robert and Henry were the more practical rubber goods dealers. The brothers saw a potential profit in providing rubber-lined suits for Taylor, and they had accompanied the two Taylors on their expedition to Florida.[2]

Taylor arrived in Buffalo in the first days of May, 1842, before regularly scheduled steamer traffic resumed for the season, but when navigation in the waters around Buffalo was clear. He hired the largest steamboat on the lake, the *Wisconsin*, to take a party of spectators out on the water to witness a demonstration of submarine armor and explosive torpedoes. The *Wisconsin* steamed to Point Albino, about twelve miles from Buffalo, and anchored in shallows about twenty feet deep. So that he could stay aboard and offer a running commentary, Taylor had recruited a sailor named Christy, a mate of the brig *Rocky Mountains*, to make the dive in the submarine armor. A description of the suit reveals that Taylor had replaced the awkward plate metal hoops on the lower torso and legs with flexible coils of copper, over which rubber trousers were pulled: "It is composed of a copper case for the head and shoulders, with india rubber arms attached. In front [of the helmet] is a small glass about three inches in diameter, to enable the operator to distinguish objects in the water. The lower part of the body and legs are also encased in a flexible copper case, giving every freedom to the limbs; over these is drawn a huge pair of india rubber trowsers with shoes attached, which is fastened to the upper case, and the wearer is thus completely protected from the water."[3]

Following the demonstration, George W. Taylor lingered in Buffalo for over a month, making inquiries about the wreck of the *Erie*. Though others believed the *Erie* had little salvage value, Taylor judged that diving work on the most infamous shipwreck in America would pay for itself in publicity for his wares and services. In the first week of June, Taylor went to Silver Creek, New York, with Captain George Miles of the steamer *Star* and interviewed witnesses to the sinking of the *Erie*. Their accounts directed Taylor and Miles to the general area in which the wreck might lay. After days of dragging lines, they found a wreck and attempted to send down a diver to ascertain if the engine could be lifted.[4]

Either the wreck was not the *Erie*, or diving to it proved to be beyond the limit of Taylor's submarine armor. No definitive word of their findings appeared in print, but two weeks later, in the first week of July, Taylor resumed his tour of diving demonstrations in Cleveland. The *Cleveland Evening Herald* mentioned that Taylor would continue on to Detroit to give another exhibition, and then would return to resume his search for the *Erie*.[5] The implication of the *Herald* article was that the *Erie* site still eluded discovery. However, once he reached Detroit, Taylor continued on northward to Lake Huron and did not return to the *Erie* for the rest of the season.

Meanwhile, in Buffalo, a man named Hammond from Fonda, New York, announced that he had discovered the wreck of the *Erie*. Hammond further made it known that he had sent a message to Taylor in Detroit, but was informed that Taylor had already departed for Lake

Huron. Hammond's bold claim was questionable, because he explained that the site had been located with the assistance of a clairvoyant.[6] Hammond's startling revelation serves as a reminder that in the 1840s, western New York State, including the Lake Erie shoreline, was in the midst of a spiritual revival that resulted in the formation of several religious movements. Collectively, this place was later named the "Burned-Over District," due to the viral spread of these new sects and "isms." Among these movements was spiritualism, whose adherents claimed that it was possible to communicate with the spirits of the dead. Mr. Hammond's clairvoyant was allegedly able to pinpoint the wreck location with guidance from the restless souls of the *Erie*'s drowning victims. Unable to find a diver to confirm the wreck's location, Hammond soon disappeared, vowing to return when the lake reopened to shipping in 1843.[7]

George W. Taylor still held an ambition to salvage a wreck in the Great Lakes region before the water surface froze solid. In early December 1842, a small steamboat—coincidentally named the *Erie*, but often referred to as the *Little Erie* to distinguish it from its tragic namesake—was trapped and then crushed by ice in Lake St. Clair, just above Detroit. It sank in shallow waters near Belvidere Bay; no lives were lost, but the passengers were forced to walk over the ice and through five miles of swamp to get to solid ground. Reports claimed that there was little or no freight aboard, and the vessel itself was valued at a modest $3,000.[8] George W. Taylor imprudently decided that the *Little Erie* should be raised, and that it should be done as soon as possible, despite the onset of winter.

Taylor's decision was a horrible mistake. He took a ship and crew of eight men to Lake St. Clair in late December. There were ice floes on the water, but the surface had not yet been closed by ice. After several attempts, Taylor, working in his diving armor in freezing temperatures, was finally able to slip chains on the *Little Erie* and raise it to the surface. At that point the waters around the wrecking crew's ship froze fast, and both it and the *Little Erie* were immobilized. Fearing for the loss of their vessel, the nine men decided to walk across the frozen lake to shore—a distance of two miles. They began their trek, but before reaching land, the ice surface parted in front of them. Taylor urged everyone to return to the ship, but four of the crew thought their chances would be better if they found a path of ice that still touched the shore. Taylor and four others returned to the ship; the next morning the ice broke up and they were able to return to port. The four men who braved the ice floes were not found.[9] The experience scarred Taylor; he never returned to the Great Lakes.

Once the ice came in that winter of 1842–1843, it lingered on the Great Lakes. Three months after Taylor's nightmarish outing to Lake St. Clair, two men hiked several miles out onto the ice of Lake Erie from Silver Creek, New York. One of the men was Orrin McCluer, a hardware goods dealer from nearby Fredonia, New York. The other man was Daniel D. Chapin, a twenty-nine-year-old self-taught mineralogist from Phillipsburg, New Jersey. Both men carried packs containing ice augers, buoys that could be placed under the ice, drop lines with soft lead weights, and special instruments belonging to Chapin.

The landscape they trekked through could have been mistaken for the Arctic. Lake Erie does not freeze to glassy smoothness. It freezes, then cracks into floes that crash into one another and refreeze. It often appears that the waves themselves freeze in place. From many vantage points, the mounds and dips obscure the view of the shoreline, so it does indeed look as featureless as the North Pole. After hiking a couple of hours, the two men halted. Chapin took pieces of a mysterious instrument from his pack and laid them out: several thin cylinders made of different metals, each about nine inches in length and shaped on one end like a skyrocket; a handle with a freely rotating spindle facing upward; and a long malacca cane, thinner than a walking stick. Chapin selected the iron cylinder and inserted the malacca stick into one end. He then affixed the cylinder horizontally to the spindle on top of the handle. Holding the handle with his arm straight out, Chapin allowed the stick to rotate freely on a horizontal plane. In operation, it looked like he was holding a huge compass needle—and Chapin appropriately named the device his "marine compass."

With the iron cylinder affixed, the instrument was tuned—supposedly—to pivot and point to the strongest, closest deposit of iron. Chapin let the giant needle lead their steps before finally settling over one spot on the ice. They used an augur to drill a hole through the ice, and into that opening they dropped the line with the soft lead weight. It is not known how many misses the men had, or even if they needed to repeat their expedition out onto the ice on multiple days. However, they eventually hit an object with their drop line that left a sharp, unnatural dent in the soft lead, and that they judged to be several feet above the lake bed. Chapin switched his marine compass to employ the gold-seeking cylinder, and found that it pointed to the same location. They marked the spot with buoys set under the ice and returned to Silver Creek. They then announced to the world that they had found the *Erie*—although that may have been more a result of trial and error than of the marine compass—and had laid buoys over the site in support of their salvage claim.[10]

Despite their discovery, McCluer and Chapin had a problem: George W. Taylor was not available, and there was no one else with submarine armor; there also were no wrecking barges with diving bells operating on Lake Erie. However, McCluer's hometown of Fredonia had a foundry. In May of 1843, the two men further announced that they were casting their own diving bell, which would weigh over four tons and be capable of descending to a depth of eighty feet.[11] The bell was completed two months later. It is not known which brave soul first volunteered to descend in the bell to the *Erie* wreck, but someone was able to reach the wreck and work free sections of the engine, which were successfully raised to the surface.

The *Cleveland Herald* reported on their limited success, noting that the bulk they brought up did not have enough value to cover the expense of its recovery. It appears that either lack of funding or difficulties in using the diving bell caused the 1843 salvage work on the *Erie* to be halted. The *Herald* did credit Chapin for finding the wreck using his ingenious equipment, and (ironic to readers today) contrasted his scientific approach to the sham efforts of clairvoyants.[12] Orrin McCluer probably bore the brunt of the cost of the 1843 effort, and may have been the unnamed commentator who later declared that he had been in the diving bell and had seen nothing of value that could be salvaged from the *Erie*.[13]

Daniel Chapin, on the other hand, was determined to find the rumored stores of specie carried by the *Erie's* immigrant passengers, who were traveling with their life savings. Chapin recruited new investors in the enterprise and kept possession of the diving bell. In March of 1844, he declared his intentions by placing new buoys over the wrecks of the *Erie* and also of the schooner *Young Lyon*, which sank in 1836 near Erie, Pennsylvania, with a cargo of rail track iron.[14] By July, three ships were stationed over the *Erie*: the brig *Toledo*, the steamer *Governor Marcy*, and the barque *Sandusky*. Using Chapin's diving bell, the wreckers were able to bring up the *Erie's* 450-foot anchor cable. After several more weeks of work, in the last days of August, 1844, they succeeded in placing chains on the wreck and started to slowly hoist it up. They had gotten the entire hull eight feet off the lake bottom when a gale hit. The *Erie* hulk had to be dropped to allow the three wrecking ships to race to port.[15] Though some work continued until October, the ship captains tallied their losses and determined they could not continue the project. The *Buffalo Commercial Advertiser* proclaimed of the *Erie*, "She cannot be raised."[16]

Chapin launched his third, last attempt to raise the *Erie* a year later, in August, 1845. The brig *Illinois* was employed to lower the diving bell; several heavy pieces of the engine were brought up. The steamer *Indian Queen* joined the work to raise the hull. After sweep chains were attached around the hull and it was raised eight feet off its bed (as far as it had been raised the previous year), the *Indian Queen* collided with the hoist pulley that was extended from the *Illinois*, causing the sweeps to break loose. The *Erie* settled once more to its resting place. The *Cleveland Plain Dealer* put into words what every local person believed: "Bad luck seems to attend everything connected with this ill-fated vessel."[17] After three years of failure, Chapin found himself a pariah when it came to soliciting new partners to raise the *Erie*. For the next half-dozen years, the *Erie* rested undisturbed, the subject of yarns told by Great Lakes mariners and dockhands of the "jinx ship."

Chapter Three

≈

End of the Taylors (1840s–1850)

For six years following Daniel Chapin's last attempt to raise the *Erie* in 1845, no diving activity of note took place on the Great Lakes. During that time, the franchise on diving activity in America that originated with William H. Taylor and continued by George W. Taylor began to fracture. These years represented the period when the fevered pace of the Industrial Revolution reached its peak. Nearly all engineering technologies underwent leaps of innovation, and the Western world spawned a generation of ingenious engineers. Therefore it is hardly surprising that George W. Taylor was joined by others in offering various approaches to submarine exploration.

One man absent from this march of progress was George W. Taylor's former mentor, William Taylor. W. H. Taylor had last been heard of in the United States in 1839, the year when his preoccupation with a battery-powered electric engine caused him to migrate to England. In 1840, Taylor displayed his model engine in several engineering exhibitions, winning monetary prizes and acclaim. However, he soon realized that the voltaic cells of that period could not be scaled large enough to offer a practical alternative to steam power. Regardless, Taylor opted to remain in England to pursue another patented idea, the mechanized cutting of steamed wood. With just a small two-horsepower engine, Taylor's saws easily ran through steam-softened wood and could fabricate 240 barrel staves per minute. It was hailed as a revolution in coopering, a craft that had changed little in hundreds of years. On hearing news from London about Taylor, a Boston paper noted: "At the present moment the worthy Capt. is all the rage in town, and he is also, as might be imagined, playing the very deuce among the coopers of England."[1]

Had Taylor kept an eye on his cooperage operation, he might have made a success of it. However, the siren call of sunken treasure reached his ears, and barely a year after starting the large, heavily capitalized stave-cutting business, he took a leave to hunt for a fabled treasure ship of the French Revolution. The barge *Telemaque* sank in the Seine on January 3, 1790, near Quillebeuf, France, almost at the mouth of the river. It was rumored to be carrying jewelry belonging to Marie Antoinette, the royal family, and other nobles, and it was laden with precious metal objects from many wealthy churches and abbeys. Decades after 1790, some claimed that there had been a secret plan by Louis XVI to have the treasures melted down at sea to prevent their confiscation by revolutionaries. The total value of the loot, estimated by William H. Taylor in 1842, was between thirty and eighty million francs. Taylor presented testimony tracing the fateful last route of the riches from several witnesses in a twelve-page prospectus titled *Salvage of the Treasure Ship Le Télémaque*.[2] Critics later said that Taylor ignored contrary evidence indicating that the treasure had been offloaded to another ship before the *Telemaque* sank. However, this second-guessing came too late, after Taylor had secured funding from investors in both England and France. With their support, in the fall of 1842, Taylor took a crew of thirty English salvage workers across the Channel to France.

The position of the *Telemaque* wreck tested Taylor's ingenuity. At ebb tide, the hulk was only eight feet under water. At that depth, even free divers could have gotten the chains around the hull. Taylor's first approach was to fasten the sweep chains to a barge on the surface, and then to let the incoming tide raise the barge and thus the wreck. However, the wreck was too entrenched in the muck, and the tide came in too fast. Taylor's barge was swamped and would have sunk if the chains holding the wreck had not snapped or been cast off.

Next, Taylor decided to drive piles all around the *Telemaque* and erect a truss work directly over the wreck. Screw jacks were then fixed to the new framework so that they could be turned to slowly lift the sweep chains attached to the wreck.[3] All appeared to be going according to plan in late November, 1842, when Taylor decided to call a stop to the work. He stated that work had to halt for the winter because the water was too cold for his workers and an unexpected volume of sand had poured down the Seine into the wreck.[4]

Rumors quickly circulated that other factors played a role in suspending the project. One report said that several casks had already been extracted and contained nothing more than oil and spoiled tallow. It was also brought to light that Taylor's laborers had received a piece of script instead of cash for their most recent fortnight of work, and that their board bills were unpaid. Taylor tried desperately to smooth things over, but when news accounts suggesting a boondoggle appeared in the London papers, his backers became alarmed. Taylor was tracked down in northern France and thrown into prison at Pont-Audemer for the debts owed to the hoteliers.[5]

The inhabitants of Le Havre, at least those who had not lost money on the *Telemaque* venture, found the fiasco to be a perfect opportunity to ridicule the English. During Le Havre's traditional carnival of Mi-Carême in March of 1843, a procession through the city streets included a float built to resemble the hull of the *Telemaque* with Taylor's truss-work structure over it. Men wearing red-haired wigs (symbolizing typical Englishmen) and masks with yellow faces and long noses cranked the float's chains in imitation of Taylor's workmen. One masquerader—representing Taylor—then stepped forward and stuck a harpoon into the interior of the fake hull, and pulled out an old iron boiler. Repeating the action, he pulled out a bunch of carrots (a play on the expression *tirer une carotte*, "to pull a carrot," meaning to trick or swindle). Finally, the actor portraying Taylor reached in and pulled out a jar of pickles. This, too, was a bit of colloquial wordplay: *Quel cornichon!* What a dick![6]

Taylor faced ten years in French debtor's prison, but was released after about a year. His mass-production barrel-making operation had barely survived during his absence, but Taylor did not have the patience to rededicate himself to such a pedestrian occupation. In 1845, he patented a new ship-propeller design. Ship propellers were fairly new in the 1840s, and Taylor's patent addressed a problem with their use on narrow canals—most propellers created a large wake that lapped at the banks of canals. Taylor invented a propeller that was said to produce no swells.

There is no evidence that Taylor's test of a small experimental propeller on England's Grand Junction Canal in October, 1845, resulted in orders or any other kind of remuneration.[7] He was declared bankrupt later that same month, and his cooperage closed.[8] Whether he was thrown into debtor's prison a second time is not known, but it was an old friend from Taylor's privateer days who finally offered him the opportunity to escape England. In August, 1847, the American naval vessel, the USS *Macedonian*, docked in western Scotland on a mercy mission. The *Macedonian* carried a relief shipment of food and grain for the starving hundreds of thousands in Ireland and Scotland affected by the potato famine. The supplies were paid for by private American citizens; the naval ship was offered by Congress and captained by a volunteer, George DeKay. This was the same Commodore DeKay to whom William H. Taylor reported in 1828, when both were in service to the United Provinces of Río de la Plata as privateers.

DeKay, who had had some crewmen fall ill on the voyage to Great Britain, made Taylor second in command on the return trip to America. After docking in Boston, Taylor travelled to Baltimore, where he intended to reestablish his diving enterprise. It is not known whether his aim was to reclaim his franchise from George W. Taylor or to compete against his former junior partner. Before any plans could be realized, William Hannis Taylor fell ill and passed away on the last day of June, 1848.[9] The "father of American diving" died at forty-two years of age, and left behind a destitute wife and child.

～

George W. Taylor may not have been as reckless in business as his one-time senior partner, but he still vacillated from one venture to another. Following his disastrous foray to the Great Lakes, where he lost four men on the ice on Lake St. Clair, Taylor returned to New York City. While there, Taylor made his first dives to the wreck of the British frigate HMS *Hussar*, a Revolutionary War–era ship. In 1780, the *Hussar* was allegedly carrying a $20,000,000 payroll strongbox when her commander made an ill-advised decision to reposition the ship from the Upper Bay of the Hudson to Long Island Sound via the East River. The *Hussar* lost steering in the strong tidal currents in Hell Gate, the section of the East River between Ward's Island (now part of Randall's Island) and Astoria, Queens. Helpless, the frigate struck a notorious hazard named Pot Rock and quickly sank. In the sixty years following the wreck, several tentative attempts had been made to recover the payroll strongbox, but all had failed. In July of 1843, Taylor and his assistants made a series of difficult dives to the *Hussar* and extracted a few objects: deck boards, copper sheathing, pieces of ballast, and a cannonball.[10] However, they were unable to progress through the turbulent waters further into the wreck.

With no other salvage work in the offing, Taylor conducted another demonstration at Castle Garden on the southern tip of Manhattan. The spectacle of blowing up ships had proven so popular that the event charged twenty-five cents per person for tickets.[11] Taylor's growing emphasis on his torpedoes in these public shows was intended to impress federal officials, especially those at the Navy Department. Taylor was trying to market his underwater explosives as a harbor defense measure, an issue that was a high priority owing to continuing tensions with Great Britain.

Taylor had a competitor in the torpedo arena: Samuel Colt. In the early 1840s, Colt was hardly known and only marginally successful. Following a disappointing attempt to mass-produce revolving cartridge firearms in Patterson, New Jersey, he had turned his attention to the design of waterproof fuses and electrical detonators.[12] Both Taylor and Colt faced resistance from some policymakers, including John Quincy Adams, who believed the very concept of torpedoes to be antithetical to standards of civilized warfare. In their view, floating defensive stationary mines that enemies were intent to run into were barely acceptable, but torpedoes that were submerged or that moved toward a target in secrecy, with no warning, were diabolical.

Perhaps due to that resistance, starting in 1845 Taylor cut back his exhibitions and concentrated on wrecking projects. To that end, he purchased a ship that the Navy had seized in an effort to curtail the slave trade.[13] A squadron of the U.S. Navy, the African Slave Trade Patrol, had been operating for decades, but had largely been ineffective. The *Spitfire* was the only slave ship captured by the United States in 1845, while at the same time Great Britain seized dozens of vessels involved in the heinous trade. Shortly after acquiring the *Spitfire*, Taylor was engaged by the Boston Mutual Insurance Company to recover 190 tons of bloom iron from the brig *Canton*, which sank in the Chesapeake Bay in September of 1845.

While working on the *Canton* wreck, an incident occurred that Taylor later related to newspapers. One his divers entered a companionway on the wreck that led down to a lower deck. After searching around on that lower deck, the diver turned to go back up the companionway and found his path blocked by a dead sailor. The dead body's arms waved outstretched toward the diver, who was horror-struck. Finally, working up his courage, he shoved the body upward, and it bobbed up to the surface. The shaken diver immediately surfaced, and was unable to sleep that night.[14] In future years, stories like this transformed the popular image of diving from a curious stunt into a romantic encounter within a different world, one full of unique wonders, dangers—and horrors.

Though Taylor had made little progress interesting the federal government in torpedoes, he did get the Navy to purchase a few sets of his submarine armor. In addition, in 1846, the U.S. government had an urgent problem that perhaps only George W. Taylor could solve: three years earlier, the first steam-powered Navy warship, the USS *Missouri*, had caught fire while on her first trans-Atlantic mission. While anchored near Gibraltar, a crewman broke a demijohn full of turpentine, which immediately burst into flames—the same type of accident that had befallen the passenger steamer *Erie* two years earlier. The crew was able to abandon ship, but the vessel burned down to its waterline and then sank just outside the harbor. Over the next few years, sandbars drifted over the wreck, rendering it a hazard to navigation. British divers inspected the wreck and declared that raising it would not be possible. The British government then began putting pressure on the United States to clean up the problem it had created.

In April of 1846, the Navy contracted with Taylor to produce a survey of the *Missouri* wreck and a report on possible salvage alternatives and their costs. Though some news accounts reported that Taylor was directly tasked with raising or removing the hulk, the only charge he was given was to lay out the options.[15] In tackling this problem, Taylor considered the possibility of using a new invention of his: large inflatable rubber bladders.[16] If several of these were attached to a sunken ship while deflated and then filled with air, their lifting power might raise a ship without the need for leverage from the surface. Taylor named the devices "camels."

Before he could go to Gibraltar and conduct a thorough survey, yet another federal imperative came to the forefront: a war with Mexico, precipitated by the annexation of Texas by the United States in 1845. In the fall of 1846, intelligence from Commodore Matthew Perry's first battle on the Tabasco River alluded to difficulty navigating past the river's sandbars. George W. Taylor offered to bring his ship *Spitfire* to the Gulf of Mexico to use his camels to lift stranded warships off the sandbars. The *Spitfire* was among the "Mosquito Fleet" ships that Commodore Perry moved up the Tabasco River in June of 1847.[17]

No accounts indicate whether Taylor's camels were needed or used with success, but regardless, Perry's fleet accomplished its goal of securing all the Mexican ports on the Gulf Coast. George W. Taylor and the *Spitfire* returned to the United States long before the war ended in March of 1848.[18] While accompanying Perry's fleet, Taylor conceived of another way in which his camels could serve ships in distress. Taylor reasoned that even smaller camels would have enough lifting power to convey a ship's anchor and chain away from the vessel, and that they could then be remotely deflated to drop the anchor at a distance, a maneuver that would be desirable in certain wind conditions. Taylor demonstrated this application for Navy officials in Charlestown, Massachusetts, in July of 1848.[19]

With his military services no longer needed, George W. Taylor resumed plans for his two ongoing projects: the salvage survey of the USS *Missouri* in Gibraltar and the tantalizing treasure hunt for the strongbox of the HMS *Hussar* at Hell Gate in New York City. In June of 1849, Taylor hired two assistants, James A. Whipple and Edward Robinson, to conduct the *Missouri* survey. Taylor relied on Whipple and Robinson, both of whom had engineering training as well as diving experience, to make the descents to the wreck, take measurements, and make technical drawings. Taylor, although he accompanied the team to Gibraltar, was unable to share much of the workload. He was confined to bed with a serious illness.

Both Whipple and Robinson, although respectful of Taylor's accomplishments, believed they had engineering skills that Taylor lacked. The survey itself was a small contract, but if Congress accepted the report and moved forward with the actual salvage—a project that might take a couple of years—there was a prospect of much larger remuneration. Whipple and Robinson fretted that Taylor's ill health might cause the government to lose confidence in proceeding with the salvage, even though they believed themselves capable, with or without

Taylor. Taylor was able to summon enough strength to return to the United States from Spain, but his future health remained in question.[20] As it turned out, fiscal constraints caused Congress to lose the sense of urgency it originally had in resolving the problem of the sunken USS *Missouri*. Taylor's report was accepted, but no immediate action was taken to award the USS *Missouri* salvage contract. Robinson, who had confidence in his own lobbying efforts, believed that he and Whipple were in a strong position to get the project even if Taylor was indisposed. Taylor, though ill, did not bide his time waiting for a decision from Congress. He turned his attention back to the wreck of the *Hussar* and decided to rely on the aid of another long-time assistant, Charles B. Pratt.

Pratt was George Taylor's most experienced diver. In 1839, he had run away from his broken home and happened to be in Rochester, New York, when Taylor was there for a demonstration of the submarine armor. Taylor unexpectedly lost an assistant, and Pratt volunteered to take his place. From that point on, Pratt often traveled with Taylor to work on his underwater demonstrations and wreck projects.[21] In 1844, at age nineteen, Pratt was married and fathered a son, whom he named Charles Taylor Pratt in honor of his employer and mentor.[22]

In the spring of 1850, Pratt visited George W. Taylor as he lay sick in his home in Washington, D.C. Taylor urged Pratt to keep working on recovery of the *Hussar* treasure. Taylor never recovered; he died on April 28, 1850, at age forty-three and was buried in Washington's Congressional Cemetery. His obituary in Washington newspapers gave a lengthy list of his accomplishments, but no details of his birth or family background. The executor of his estate was Major William Scott, a procurement agent for the Navy Department with whom Taylor had often worked.[23]

There were bitter feelings over the disposition of Taylor's estate. Whipple and Robinson, who performed the USS *Missouri* survey work for Taylor in Gibraltar, had never received their share of the survey contract, and so made a claim against the estate. Taylor willed his diving equipment to Pratt, a gift much resented by Taylor's widow, the former Rebecca Hawkes of Lynn, Massachusetts. Mrs. Taylor also suggested to Whipple and Robinson that, should they get the federal contract to salvage the USS *Missouri*, a share of that should rightfully go to her. Whipple and Robinson judged that to be presumptuous, since Taylor had done so little of the survey work.[24]

The death of the Taylors, and with them the Taylor diving franchise, left a vacuum in American submarine operation enterprises. Charles Pratt, though a skilled diver, was neither a mariner nor an engineer. Whipple and Robinson, the most likely inheritors to George W. Taylor's work, became stuck in a kind of limbo, waiting to hear if they would receive the USS *Missouri* salvage project. They were therefore unable to commit to any other large, long-term salvage or treasure-seeking ventures. Whipple even resumed his employment as an engineer at a locomotive works in Boston. The Goodyear brothers and other rubber companies continued to sell submarine armor outfits to sales outlets across the country.[25] They limited their activity to manufacture and did not evidence interest in further research and refinement of the diving dress.

However, there was no shortage of diving entrepreneurs ready to step in to fill the gap left by George W. Taylor. Moreover, to many it was not even clear that submarine armor was the best technology for every underwater engineering need. Other inventions, techniques, and technological advances were under development in 1850 that could be used to dislodge undersea obstacles, raise wrecks, recover sunken cargo, erect submerged foundations, and conduct military operations under the waves. Not only was it not obvious which technology was best suited for a given underwater job, it was not known which was less dangerous to its human operators.

Chapter Four

≈

The Marine Engineers (1840s–1852)

In the first years of the 1850s, many of the best marine engineers working in America were poised to apply their talents to salvaging wrecks in the Great Lakes. Before describing the specific circumstances that turned their interest to Lake Erie, a brief sketch of each of these individuals will provide context for those later events.

The Demolitionist: Benjamin Maillefert

Many times the objective of a wrecking operation is not to extract cargo from a sunken hulk or to recover any part of the ship structure or engine. Instead, the goal is merely to remove it as an obstruction to navigation. In this respect a shipwreck is no different than a hazardous shoal or a protruding outcrop of rocks. The first use of explosives to clear harbors, channels, and rivers of such hazards is not documented, but it probably did not occur until the early nineteenth century. In the 1830s, British military engineer General Sir Charles William Pasley used gunpowder to clear wrecks out of the Thames River. In 1840, Pasley used explosives to break up the wreck of the *Royal George*, after divers using Deane and Siebe equipment had extracted its guns and other articles.

In 1847, an Anglo-French engineer named Benjamin Maillefert was residing in Nassau, Bahamas, when a merchant ship sank in the city harbor, threatening shipping traffic. Maillefert first tried setting off eighty charges of black powder under the submerged hull. These had only a small effect on the wooden structure, but Maillefert noticed that the rocks beneath where the charges had been set had shattered. He correctly surmised that it was the concussive power of the water that had crushed the rock. To test his theory, he placed a charge on the upper portion of the wreck, which when detonated completely tore the timbers into splinters.

Maillefert next applied the technique of concussive blasting to Rockfish Shoal, a coral reef at the mouth of Nassau harbor. He removed 900 tons of rock and deepened the passage from eleven feet to eighteen feet, making it safe for most ships. Bolstered by success, Maillefert headed to the United States to offer his services in removing the most notorious navigation hazards in the Western Hemisphere, the rocks at New York's Hell Gate. After presenting his offer in New York City, Maillefert immediately headed up the coast and made a similar offer to remove the Southwest Ledge reef from the harbor of New Haven, Connecticut. However, when he went out and took soundings of Southwest Ledge, he found it to be too immense. It was later decided to site a lighthouse on that reef rather than attempt its removal.

Federal, state, and city governments balked at funding the clearing of Hell Gate, so a committee of the wealthiest shipping merchants in New York, led by Henry Grinnell, finally stepped in and contracted Maillefert to remove the most dangerous single obstruction in Hell Gate, Pot Rock. In spite of widespread skepticism about his chances of success, Maillefert began operations in June, 1851.[1] During the first week of work, his float platform, anchored near Pot Rock, was hit three times by vessels caught in the whirlpools. The tides allowed him just minutes per day to set and detonate the charges, but progress was visible each week. His assistants would row out and lower the charges, leaving them suspended above the rocks, and after they rowed clear,

Maillefert would trigger them using an electric battery. The charges consisted of loads of black powder; the development of dynamite and TNT was still more than a decade away. Navigation around the Pot Rock area improved, but Maillefert's contract specified that the rocks had to be cleared to a certain depth. His work on Hell Gate would continue for several years, but by early 1852 New York's business leaders were convinced that Benjamin Maillefert was the nation's most ingenious marine engineer.

The Untethered Diving Bell: Henry Beaufort Sears

In early 1852, two young men, Edgar W. Foreman and Henry Beaufort Sears, worked together to construct a model that Foreman had designed of a sophisticated new type of diving bell. Foreman described himself as an architect, but, lacking patrons, he worked in New York City as an artist and draftsman. Sears was a West Point graduate and veteran officer of the Mexican-American War. He had spent the years 1850–1851 in what his résumé described as "explorations in South America." The exact nature of that journey is not known, but coincidentally, during the early 1850s, several small intelligence-gathering and survey expeditions in Central and South American were sponsored by the United States government. That experience may have brought Foreman and Sears together and spurred their diving bell project. It appears that they, like the Taylors and many others, believed that tropical pearl harvesting was the path to easy fortune.

Diving bells had several limitations: they had to be heavy to descend counter to their internal air pressure, and therefore they needed a strong hoist mechanism to suspend the bell over a site and to lift it out of the water; if the support ship rode on rough seas, the diving bell would be tossed from side to side and up and down, making work impossible. Some diving bells had glass portals, but visibility was limited. Also, although in some cases free divers exited and reentered bells, for the most part they were used for work directly over a hatch in the bottom of the bell. On top of everything else, they were expensive to forge, requiring a large amount of metal.

Foreman and Sears sought to eliminate many of these problems by equipping their diving bell with ballast chambers and compressed-air tanks, so that the divers within the bell could control its ability to descend and ascend by filling the tanks with air or water. Also, they added a screw propeller on one side that could be used to move the bell laterally. Their "Nautilus" bell still needed air-hose connections to the surface, but at least the bell could move around over a wide work area without constant repositioning of the ship above. The bell could also be moved against a current by deploying and then cranking in its own anchor chain.

On July 8, 1852, while Foreman and Sears were in the midst of building their first test model, Foreman took a break and went swimming with a group of friends in Long Island Sound near his New Rochelle home. He went out farther into the water than his companions, and then disappeared. His drowned body was later recovered. Henry Sears, though not the original inventor of the Nautilus bell, hired machinists to test and further perfect and improve the design. Sears filed a patent application that named Foreman and his executor, his father Jonathan Foreman, as the patentees and Sears as the assignor. Despite the loss of Foreman, by late summer, 1852, Sears had a working model of the Nautilus bell and was anxious for an opportunity to demonstrate its potential to the public.

The Submariner: Lodner D. Phillips

The concept of vessels that can move under the water goes back centuries. Since the 1700s, designers from various countries using differing concepts had made many attempts to realize that idea. Most of these early submarines were unsuccessful, due to bad luck, lack of financial support, and a number of technical obstacles. Among those challenges were depletion of oxygen, a means of propulsion, ballast systems, lateral and horizontal stability, watertight seals, visibility, and manipulating objects outside the vessel. These problems appeared so insurmountable that few established shipbuilders or naval departments wasted time considering submarine design. It was an endeavor left to visionary inventors and tinkerers, working with no margin for error.

Two of these submarine pioneers were Americans: David Bushnell and Robert Fulton. Bushnell's contributions to underwater engineering were multiple: he proved that gunpowder could be detonated underwater, and he built a one-man submersible, the *Turtle*, to attach mines to British ships during the Revolutionary War. The *Turtle's* drill, intended for use in attaching mines, could not penetrate the copper plating under the British warships. The *Turtle* sank while retreating from an engagement, though its pilot escaped without injury.

Fulton, while researching canal design on a visit to the French First Republic, became interested in continuing Bushnell's efforts. He was able to convince the French to support the construction of a twenty-five-foot ship that could dive to a depth of twenty-five feet. The initial tests of Fulton's *Nautilus* (not to be confused with Sears's later Nautilus diving bells) were successful enough to warrant approval of new all-brass models. However, Napoleon grew distrustful after Fulton dismantled the first *Nautilus*, thinking that Fulton did so to hide his engineering secrets. The cold-shoulder Fulton received from the French, perhaps combined with Fulton's disillusionment with the way Napoleon was leading France away from republican ideals, caused Fulton to take his ideas to France's foremost enemy, Great Britain. Admiral Nelson's decisive naval victory at the Battle of Trafalgar caused the British in turn to cool their interest in Fulton's submarine. Disgusted with European powers, Fulton returned to America and later achieved great fame with steamships—and never mentioned submarines again.

For several decades after Fulton, there was a gap in American interest in submarines. In the 1840s and early 1850s, only one American was known to have been researching submarine vessels: a shoemaker from Michigan City, Indiana, named Lodner Phillips. The secretive Phillips left no notes and no explanation for what sparked his interest in diving boats. He came from a long line of shoemakers, so one motivation may have simply been to break free of that predestined mold. At any rate, Phillips's cobbler skills were related to one of the technical challenges to building a submarine—the need for waterproof seals for several of the vessel's features, including hatches, the propulsion mechanism, and external tools manipulated from the inside. Prior to the introduction of rubber, leather was used for this purpose; indeed, the very first diving suits were made of leather, not rubber.

Phillips built his first submarine craft in 1845, when he was twenty years old. Few details are known about the appearance of his first model, but it was described as being shaped like a whitefish (native to nearby Lake Michigan) and covered with sheet copper. For propulsion, it had a push-pole that passed through a rubber-sealed opening in the bottom of the hull. It also had a water-filled ballast tank, emptied by means of a plunger, to make the vessel dive or rise. It is not known if Phillips made any successful tests; what is documented is that the submarine sank in Michigan City's Trail Creek, in about twenty feet of water.

Sometime between 1845 and 1850, Phillips moved to Chicago, Illinois. Circumstantial evidence suggests that he built a second submarine in Chicago, and that it sank in the Chicago River. Nearly seventy years later, in November, 1915, a submarine of unknown origin and vintage was dredged up from the river. According to Phillips's nephew, the discovery was his uncle's second submarine model. The curiosity was dubbed the "Fool Killer," and it was put on display in dime museums and traveling shows, complete with a concocted, fanciful history of its origins. Unfortunately, the artifact disappeared from public view in the summer of 1916, and its fate remains a mystery. Photographs of the vessel taken at the time it was raised display features identical to those found in Lodner Phillips's few drawings. With no other plausible candidates, it appears very likely that the Chicago River submarine was built by Phillips.[2]

Phillips returned to Michigan City to construct his next submarine. Lacking funds, he relied on loans cosigned by his father and two brothers. Phillips's signature on the notes was written backward, perhaps in homage to da Vinci, a minor detail that adds to his curious character. More specifics are known about this craft's design, because Phillips submitted a patent in 1852 for its steering mechanism—a hand-cranked propeller mounted on a universal ball joint so that the shaft could be pushed at different angles to shift direction.

This vessel, nicknamed the *Marine Cigar*, was about thirty feet long and just four feet wide. In addition to the propeller mounted on a ball joint at the rear, at the prow was a similar ball-and-socket joint for manipulating tools on the outside of the vessel. The side length of the body included several round glass portals, with another portal on the top surface. Drawings of the craft that appeared in *Scientific American* also indicated that a hatch existed in the bow allowing egress for a diver. Later in the 1850s, Lodner patented a solid-metal atmospheric diving suit designed for this use.

No contemporary accounts exist of the testing of this submarine, but years later, Lt. Francis Morgan Barber of the U.S. Navy mentioned that Phillips had taken his wife and two children on a day-long voyage on Lake Michigan. A crucial feature was an air-scrubbing mechanism to eliminate carbon dioxide, and compressed air canisters to provide fresh air. Longitudinal stability was achieved through fore and aft water reservoirs with valves activated by a pendulum weight.

As fantastic and ahead of its time as it was, few people outside of Phillips's closest family and friends were aware of his creation. Out of either fear of failure or protectiveness of his patents, he shunned public demonstrations of the vessel in operation and did not speak to newspaper reporters. It appears that Phillips did not want his invention unveiled until it had proven itself in practical operation. In April,

1852, he wrote to Secretary of the Navy William A. Graham offering to demonstrate its capabilities after it was completed in three months' time. Graham referred the letter to the Navy's Bureau of Construction. The chief of that bureau said that he had no authority to test or purchase submarine boats, and that "the boats used by the Navy go *on* and not *under* the water;" Graham quoted these remarks in his reply to Phillips.[3]

Failing to interest the military, Lodner Philips needed a spectacular commercial demonstration of his invention. The salvage of a notable wreck would present an ideal opportunity.

The Lifters: James Eads and Albert D. Bishop

In 1833, thirteen-year-old James Buchanan Eads arrived in St. Louis in a manner that foreshadowed his later career. As the paddle steamer that he, his mother, and two sisters were passengers on neared the city's piers, the ship's chimney pipe collapsed and showered the decks with sparks, immediately igniting a fire. The panicked passengers jumped to safety as soon as the dock was within reach, but eight people were killed. Eads and his family were lucky to survive. The ship burned to the water, destroying the cargo, which included all the belongings of the family.

Despite (or because of) this experience, Eads was lured by the life of steamboat navigators, and at age nineteen took a position as a lowly clerk aboard a steamer named *Knickerbocker*. Over the next three years, Eads saw firsthand the dangers of travel on the Mississippi: the shifting course of the river, the moving sandbars, and the constant threat of submerged tree snags. The design of the boats themselves brought another set of risks: to save weight and draft, they were built with light hulls, the steam boilers were designed with few safety measures, and paddle propulsion made vessels slow to respond to commands from the wheelhouse. Eads witnessed the loss of several ships and their cargos, many in just a few feet of water.

By age twenty-two, Eads recognized the need for a service that had unlimited potential: the salvage of sunken riverboats and their cargos. He had taught himself the basics of engineering mechanics, and realized that specialized equipment was needed, primarily a vessel capable of hoisting great weights and lowering diving bells. Eads approached St. Louis shipbuilder William Nelson with a brash offer: if Nelson built a wrecking boat to Eads's specifications, Eads would offer a partnership in his salvage business. An agreement was reached.

While his new wrecking steamer was under construction, Eads contracted to recover a load of pig lead from a sunken barge near Keokuk, Iowa. For the job, he sought out George W. Taylor, then currently touring the Great Lakes area. Taylor (or one of his trained divers) joined Eads and tried to descend into the river, but the current made it impossible to stand, and the muddy water made it impossible to see. Eads improvised a diving bell by obtaining an empty forty-gallon whiskey barrel from Keokuk, knocked out one end, hung pig lead weights on it, and attached a block and tackle to the closed end. The Taylor diver refused to get inside the makeshift diving bell, so Eads did so himself. After Eads set the example, the diver finally agreed to continue the work.[4]

From that incident forward, diving bells were Eads's preferred equipment for river work. The first wrecking steamer, the *Submarine No. 1*, was a success, but the business quickly consumed Eads's time and family life. Bowing to personal concerns, he spent the years 1845–1847 off the river, managing a glass factory, but that effort failed due to poor market conditions brought on by the Mexican-American War. In 1848 Eads returned to the wrecking business with a new boat, *Submarine No. 2*, and quickly reestablished his supremacy in Mississippi River salvage. Demand for his services required the construction of a third wrecker, *Submarine No. 3*.

James Eads's fortune was secured by a cataclysm that took place on the river on May 17, 1849. The steamboat *The White Cloud* caught fire while docked at St. Louis's piers. The flames burned the ropes tying down the vessel, and as it drifted out into the river it passed other steamers and set them afire. Embers were blown from the burning ships onto shore, where waterfront buildings soon erupted in blazes. Over the next twelve hours, over 400 buildings and thirty river vessels were destroyed. The enormous job of raising and salvaging the wrecks fell to James Eads and his *Submarine* steamers. Eads poured the profits he made into the construction of his most ambitious wrecking steamer yet, *Submarine No. 4*. This vessel had two parallel hulls, with a deck placed between, offering maximum stability for lowing a diving bell through an opening in the center and for lifting hulks using a derrick that ran the length of the ship.

Mississippi River traffic provided Eads with all the business he could handle. He was doubtless aware that the Great Lakes were filling with unrecovered wrecks, but he also realized that underwater work in lakes might require different techniques and different wrecking vessels than his *Submarines* with their diving bells and sand pumps. Moreover, Eads's ingenuity sought other engineering challenges that the Mississippi offered, such as St. Louis bridge construction and structures to manage the navigability of the river. It was to these projects that Eads turned in the latter part of his career, but only after he had answered the call to serve the United States during the Civil War.

The success of James Eads was not lost on another civil engineer, Albert D. Bishop of New York. In the late 1830s, Bishop, a carpenter, began to specialize in the construction of marine and dockside cranes. The earliest instance in which he is known to have been involved in the wrecking business was in 1840, when he contracted with British officers to dislodge the *Eagle*, an American slaver brig that had sunk near the docks in New York harbor. The *Eagle* had been captured by the British brig *Buzzard* in 1839 as part of the international effort to suppress the slave trade. The incident involving the *Eagle* had been an embarrassment to the U.S. government: a British warship had entered the harbor escorting a captured American vessel. Bishop was able to refloat the *Eagle*, but not before some contentious disputes with the British over the terms of his contract.[5] The episode serves to characterize Bishop's career, which was often tracked in newspaper accounts of court proceedings over business disputes; he was sometimes the plaintiff and sometimes the defendant.

By 1846, Bishop had invented and exhibited a new derrick design that included a movable base on tracks and a long, rotating horizontal boom near the top of the tower. Attached to the boom were multiple block-and-tackle systems. The configuration was both gigantic and unique, and Bishop secured a patent. Bishop's Patent Derrick was hailed as being unsurpassed in its lifting power; he calculated that a sufficiently large design would be able to draw up a wreck weighing 1,000 tons. To be able to lift that much weight, Bishop devised alternative configurations of counterweights: on smaller models, the derrick could sit on a barge that contained water ballast compartments that could be flooded as a wreck was raised. On a larger model, the derrick barge could be physically linked to another ship on the opposite side from where the wreck was to be raised.

By 1850, Bishop had noted both Eads's success in the wrecking business on the Mississippi River as well as the growing number of ships and valuable cargos lost on Lake Erie. The apparatus for fastening chains around a sunken hull was not his primary concern; he was willing to use free divers, diving bells, or men in submarine armor. In October of 1850, from his home in Brooklyn, Bishop advertised for investors in a company to be formed for "the purpose of putting into operation one of Bishop's floating derricks on Lake Erie."[6] The company came together in April of 1851 with an impressive list of officers: Rem Lefferts, heir of a prominent Bedford, Brooklyn family; Aaron D. Patchin, president of a Buffalo bank; Watson A. Fox, a colonel in the Buffalo National Guard; Lucius H. Pratt, a leading shipping and wharf warehouse businessman of Buffalo; and Charles Manning Reed, the Erie, Pennsylvania, steamer magnate whose prize ship, the *Erie*, had burned into the water in the "awful calamity" of 1841.[7]

Bishop's new Lake Erie wrecking company got off to a promising start. In February, 1852, Bishop was able to raise the steamer

Mayflower of Ward's North Shore Line.[8] The *Mayflower's* hull was only lightly damaged, so repairs were made and it returned to service later that year. In June, the steamer *Keystone State* sprang a leak and settled in the shallows near Detroit. Bishop was able to box off a section of the hull and pump the water out, allowing repairs to be made without putting the vessel into dry dock.[9] It appeared that Bishop was poised to become the wrecking master of Lake Erie, prepared to operate on the most challenging shipwrecks.

The Commercial Diver: James A. Whipple

In 1850, James Aldrich Whipple was heir apparent to the Taylor diving operations, an enterprise that existed more in concept—involving separate contracts and projects—than as a distinct corporate entity. Whipple had worked as a diver for George W. Taylor beginning in 1847, when he was twenty-one years old. At that time, Whipple was an apprentice engineer at the Boston Locomotive Works, the leading American rail engine factory and home to some of the best engineers in the country—including Whipple's father. If there was a vanguard of the Age of Steam in America, the draftsmen of the Boston Locomotive Works were at its front. Whipple went as a spectator to one of George W. Taylor's diving exhibitions and observed that an obvious improvement could be made to the Taylor submarine armor—there was no need for a heavy, cumbersome exhaust hose, because a one-way valve could vent the spent air directly into the water.[10]

Whipple first tested his armor modifications in Jamaica Pond, Boston, with satisfactory results. While still employed in the locomotive works, he took time off over the next year to do some salvage diving off the coast of Maine. In the summer of 1849, he and Edward R. Robinson joined Taylor in Gibraltar to commence a government-contracted survey on the wreck of the USS *Missouri*. Robinson had some practical engineering experience as a blacksmith. He was also politically active on behalf of the anti-Catholic, anti-immigrant Know-Nothing Party. He may have believed that his political convictions would find a sympathetic ear with Secretary of State Daniel Webster, who would be the final arbiter in granting the *Missouri* contract. Whipple came from a more-reticent Quaker background, so it is not known if he shared Robinson's politics.

Taylor fell ill almost immediately, leaving Whipple and Robinson to conduct the dives over the USS *Missouri* and to supply the documentation for the survey report. The report was delivered to the navy in the summer of 1850. It suggested the removal of the USS *Missouri* would cost the modern equivalent of over two million dollars. The U.S. Congress took up the matter in the Naval Appropriations Bill debated in September. The price tag of the project, in addition to the death of Taylor, caused Congress to hesitate.

The delay was maddening both to Great Britain, which held Gibraltar harbor as a strategic port and needed it cleared, and to Whipple and Robinson, who were confident that they alone could tackle the job of raising the ship. The USS *Missouri* lay across the Atlantic Ocean, and it would be a long, complicated, and lucrative project. For those reasons, while the award of the salvage contract by Congress lay unresolved, Whipple and Robinson could not commit themselves to any other long-term wrecking operations.[11]

However, while he awaited action on a new appropriations bill from Congress, Whipple agreed to assist with a quick, short-term salvage contract off the coast of Long Island, near Fire Island. In July of 1850, the ship *Elizabeth*, bound for New York from Italy, struck a sandbar just a hundred yards from shore. As the ship foundered, there was sufficient time to rescue a few of the passengers, but several of the crew seemed more interested in their own safety; to add to the shame, pilferers on shore were too busy gathering goods that washed up on the surf to come to the aid of those left on board. Among those abandoned to drown was Margaret Fuller, the famous transcendentalist and women's rights advocate. Her body and that of her husband were never found, though her friend and sponsor Ralph Waldo Emerson sent Henry David Thoreau to search up and down the shoreline for her remains. On board the *Elizabeth* was a marble statue of Senator John C. Calhoun crafted by sculptor Hiram Powers. Powers, an American living in Florence, Italy, had been commissioned by the state of South Carolina to create the monument to Calhoun, one of the most famous orators in American history, who had died early in 1850. Calhoun was a fierce defender of the institution of slavery. When the ship carrying the statue was wrecked, free divers attempted to attach grapples to the box carrying the statue, which lay in water just a dozen feet deep at low tide, but it was too deeply embedded in the soft sand. Whipple was called in to use his submarine armor. He attempted the task several times, but was buffeted by strong currents. Finally, on a day when the water was more placid, he was able to attach the hooks and the statue was lifted out of the water.[12]

The rescue of the Calhoun statue was widely reported and captivated the attention of the country. Whipple had done something grander than salvaging lost specie and machinery; he had rescued a work of art and a symbol of Southern civic pride from a watery grave. Because of this feat, for a time in the early 1850s, he was the most famous diver in America. All he needed was a new project, but lacking a final decision on the USS *Missouri* contract, he returned to the Boston Locomotive Works to await developments.

The Entrepreneur: John E. Gowen

When Congress finally approved a naval appropriations bill that included funds for the removal of the USS *Missouri* wreck in Gibraltar in March of 1851, a short period was declared for sealed bids to be submitted by parties wishing to contract for the work. James A. Whipple and Edward Robinson believed their offer could not be matched in substance, nor could it be underbid. However, just a week after all bids were entered, it was announced that the contract had been awarded to another Bostonian, a man named John Emery Gowen.

Gowen, born in 1824, had less than two years of experience in diving operations and no experience in raising or removing entire ships. For a few years prior to 1849, Gowen had been in the import business in Boston with a partner, Thomas F. Wells. Nothing in the background of either man indicated familiarity with marine engineering, diving, or the wrecking business, although as importers, they must have been keenly aware of cargo losses due to shipwreck. Once they struck on the idea of recovering freight, they probably investigated where they could obtain submarine armor, and they would then have discovered that William H. Taylor's patent rights, now protected by George W. Taylor, were either unavailable or priced too high. Instead, they turned to sources in England, which had a reputation for quality diving apparatus.

Their first attempt to recover lost cargo was over the wreck of the HMS *Plumper*, which had sunk off the coast of New Brunswick in 1812. The *Plumper* carried payroll cash intended for British troops in New Brunswick; in 1850 the amount lost was claimed as equivalent to $350,000. Two previous salvage attempts using diving bells had recovered a fraction of that amount, but most of it remained undiscovered. Wells and Gowen recruited two divers from England, who brought their own equipment. If it was a concern, their use of English submarine armor outside of U.S. territorial waters would render mute the issue of infringement on Taylor's patent. The diving was done from a small twenty-foot open boat. The English divers found some coins and artifacts, a difficult task since much of the wreck had disintegrated over the years.[13] No fortune was gained through their efforts, but Wells and Gowen were encouraged enough to continue in the business.

To forestall possible patent disputes, they decided to construct their own submarine armor using the English diving equipment as a template. To do so, they sought out craftsmen in Boston, and doubtless took advice from their hired English divers. As long as their armor introduced improvements to a patented design, contemporary law would not have viewed that as an infringement. Exactly what improvements Wells and Gowen made cannot be ascertained, because the improvements to the Taylors' American apparatus—in comparison to the Deane and Siebe apparatus in England—that occurred over the previous fifteen years are not precisely known.

The two young Bostonians began to promote their brand of submarine armor even as the first examples were being assembled. Advertisements published in March, 1850, targeted one specific market: California Gold Rush prospectors. The copy promised that the gear was "furnished to be portable to carry across the Isthmus" [of Panama, a major route to California].[14] At least one prospector took up the offer and reported to newspapers a year later that he had made a small fortune from alluvial gold found on river bottoms using the Wells and Gowen diving apparatus.[15]

Lured by stories of numerous other wrecks, Gowen and his divers returned to the Bay of Fundy in the spring of 1850. The expedition recovered artifacts, machinery, ironwork, and some coinage, but again, no treasure trove.[16] Later in 1850, Gowen conducted dives on the wreck of the *Lexington* in Long Island Sound—the same wreck that George W. Taylor had visited shortly after the steamer burned and sank in 1840. Gowen's *Lexington* expedition was not much more productive than Taylor's—the strongbox containing $80,000 was not located.[17]

After receiving the award for the USS *Missouri* contract in March of 1851, John E. Gowen assembled a crew of divers and purchased a captured slave trade ship (just as George W. Taylor had done), the *Chatsworth*, from which the operations in Gibraltar would be based. One piece of gear that Gowen brought was a version of Albert D. Bishop's Patent Derrick. Gowen also brought William Irwin's inflatable camels, which were a larger version of G. W. Taylor's rubber camels. Although Irwin's camels had greater lifting power than Taylor's earlier invention, they proved ineffective against the weight of sections of the

USS *Missouri*. The Bishop's Derrick variation, on the other hand, proved instrumental in lifting tons of heavy machinery out of the water.

The contract allowed three years to complete the work of removing the USS *Missouri*. Before starting, Gowen boasted that he could do the job in six months, and he very nearly made good on that promise. Work began in August of 1851 and was completed by May 1, 1852, much to the satisfaction of the governments of the United States and Great Britain.[18] By the summer of 1852, John E. Gowen was hailed as the foremost underwater contractor in the world.

The Steamship Builder: Daniel R. Stebbins

Daniel R. Stebbins was a devoted machinist in the heyday of steam. In 1835 he moved to the Maumee City, Ohio, frontier from his Lake Ontario hometown of Sackets Harbor, New York, and there established a machine shop. Shortly after arriving in Maumee, he married and started a family. Sadly, Stebbins soon learned that mastery over iron and steam power had little bearing on overcoming the fragility of life—four of his five children died in infancy.[19]

Stebbins shop was prosperous enough to allow him to make the natural progression from working on engines to building a small steamboat. In 1844, the steamer *J. Wolcott* was built for James Wolcott, a successful Maumee businessman. The *J. Wolcott* was put into service on the short route from Maumee to Detroit.[20] With one steamboat to his credit as a shipbuilder, Stebbins drew up plans for a large passenger steamer to serve the burgeoning immigration market.

The result of his efforts was the *G. P. Griffith*, a Lake Erie palace steamer weighing 600 tons and reaching 193 feet in length. It was built in partnership with Seth P. Newell of Maumee and two merchants from Buffalo, Richard Sears and John Morris Griffith. Griffith was the son of Griffith Pritchard Griffith, a transportation magnate from Troy,

New York. By the time the steamer was ready to be put into service in 1848, Stebbins ended up owning only an eighth share, which some later claimed was due to manipulation on the part of his associates.[21] However, Stebbins agreed to serve as the onboard ship engineer, a position he enjoyed more than being majority owner of a business concern.

In the spring of 1850, the *G. P. Griffith* was sold to Charles C. Roby, a veteran steamship captain despite his youthful age of thirty-four. Stebbins agreed to stay on as chief engineer, and maintained his one-eighth share. The ship's first run of the season from Buffalo to Toledo was delayed until mid-June, due to repairs on the Erie Canal that delayed the year's first wave of immigrants. Captain Roby was justifiably proud of his luxurious side-paddle steamer, and had invited his mother, wife, and daughter to make the first trip. They boarded the *G. P. Griffith* in Buffalo on Sunday, June 16, along with 322 others including crew and passengers. Over 200 of their number were immigrants from Germany, England, and Scandinavia. A few hailed from other countries, and the remainder were Americans looking to relocate or making business or family visits.

The *G. P. Griffith* made brief stops at Erie, Pennsylvania, in the afternoon, and at Fairport Harbor in the early hours of Monday, June 17. Shortly after leaving Fairport Harbor, a little before 3:00 a.m., the helmsman noticed flames coming up around the smokestack. Crewmen attempted to douse it from above, but the fire continued to grow. The wheelsman steered for shore as Captain Roby was roused from his cabin. The passengers were wakened and a few cried in alarm, but the shore appeared to be close enough that the steamer could run aground before the fire spread.

A half-mile from safety, the *G. P. Griffith* struck and ran up onto a sandbar. Before the crew could even attempt to get off, the flames spread throughout the vessel. Once he realized that the lifeboats could not be reached, Captain Roby ordered the crew to throw piles of firewood into the lake to serve as something to grab onto. Passengers poured over the side fully clothed, some weighed down by money belts or by coins sewn into dress hems. Others ran into the flames to rescue loved ones or possessions and burned to death. Captain Roby remained on the upper deck with his family, where he could see the full horror unfold. Flames finally drove him down to the main deck. There, at the last moment, he threw his mother and child overboard, and then leapt off with his wife. The entire family quickly disappeared below the surface.

Daniel Stebbins had remained at his engine post until the ship hit the sandbar. He assisted other crew members until the last minute, then jumped into the water, thick with bobbing passengers. Stebbins knew that no lifeboats had gotten off, and had seen no other ships nearby. He judged that the only chance to save those in the water was if help came from shore. He was a strong swimmer, and made the distance quickly. Once on the beach he found a small skiff, but no people to aide him, so he put the boat in and rowed back out to the burning wreck. He pulled floating survivors aboard and headed back to the shore. Twice more he went out, saving a dozen lives. After the third trip, he collapsed on the shore in utter exhaustion.

A score of the survivors were able to swim to shore; and others were picked up by ships that eventually arrived on the scene. In all, just over 40 people out of nearly 320 survived, a worse loss of life than the *Erie* had seen nine years earlier. The harrowing tales of those who made it off the ship were published in newspapers throughout the country. The aftermath brought its own melodrama:

- A young Irish boy was escorted past the rows of recovered dead bodies, and pointed out one family member after another.

- A young German toddler was found floating on a plank, entirely dry, his name and relations unknown and presumed dead.

- The day after the disaster, searchers rowed out to the wreck, which had come to rest in just a dozen feet of water. One of the survivors, a man, accompanied them. They were able to stare down into the clear water and see the bodies strewn on the lake bed. The survivor was able to identify his wife by looking down to the bottom and recognizing her clothes.

- The makeshift coroner's shed had no means to preserve the bodies; and most of the immigrants' names were unknown. The decision was made to bury ninety-four bodies in a mass grave by the shoreline. However, as word spread that many of the foreigners had carried wealth on their bodies and clothing, concern mounted that the graves would be violated. An ad hoc committee of German residents from Cleveland heard about the mass grave and raised money to purchase a quarter-acre of land near the site; they exhumed the bodies, placed them in separate caskets, and reburied them.

- A salvage ship took the *G. P. Griffith* in tow, but after it was brought toward the mouth of the Chagrin River, the lines broke and the *Griffith* sank in shallow water.[22]

Daniel Stebbins had lost his job, his one-eighth owner's share, and the ship he had built. He had witnessed unimaginable suffering. After he collapsed on the shoreline, he never seemed to be the same man again. After he died twenty-four years later in 1874, it was noted "The event always followed Mr. Stebbins with most depressing influence, while the physical injury sustained was permanent."[23] No one questioned whether Stebbins had done his duty, and no one doubted that his heroics had saved many lives. Yet when it came to identifying the cause of the fire, some suggested that it had been a result of his actions.

Stebbins had brought five kegs of a new engine lubricant on the voyage, and was said to have used some of the grease on the *G. P. Griffith's* fateful trip. Beef tallow had been the traditional source of grease, but it contained acids that corroded iron. Stebbins probably employed a mix of tallow and mineral oils, which, in theory, was an effective combination. However, a few contemporary accounts question whether it was too flammable and might have caused the fire. Against these rumors, all the witness accounts (including Stebbins's own testimony) indicated that the flames came first from the freight hold, away from the engines. The cause of the fire, and the possibility that it might have been caused by any of his actions or inactions, haunted Daniel Stebbins long after others had ceased to wonder about it. In the months that followed the disaster, his conscience was gnawed by doubt. He had to find the answer, even if it lay on the bottom of Lake Erie.

Part Two

THE HEROIC AGE OF DIVING

Chapter Five

≈

The *City of Oswego* (July 1852)

The early 1850s represented the height of the era of the "palace steamers" on the Great Lakes. These were side-paddle steamers over 200 feet in length, designed especially for large-volume passenger traffic to serve the waves of westward movement by European immigrants, many of whom reached Lake Erie through the Erie Canal. These large steamers were too wide to pass through the Welland Canal that linked Lake Erie to Lake Ontario and the St. Lawrence River; and before 1855 no ships could surmount the St. Marys Rapids that stopped water traffic between Lake Superior and the lower Great Lakes. Even after the Soo Locks were built in 1855, the large palace steamers could not fit through them; they were restricted to Lake Erie, Lake Huron, and Lake Michigan.

In the 1840s and early 1850s, at the same time the palace steamers were built, other trends were underway that would eventually make the big side-paddle steamboats obsolete. Railroads were spreading across the northeastern United States, reaching the Ohio River by 1852. The huge engines that powered paddle wheels sacrificed hold space and made those steamers inefficient freight carriers, which was a factor that began to favor rail freight. Freight traffic on the Erie Canal began to dip, forcing its operators to widen the canal to accommodate larger and more efficient vessels. The older technology of sailing schooners, though slow, offered larger freight holds and could handle large, bulky containers better than paddle-wheelers. These trends combined to limit the heyday of the palace steamers to little more than a dozen years, between 1844 and 1857.

One additional factor that argued against side-paddle steamers was the advent of a new propulsion method for ships: screw propellers. Inventors in Great Britain had pioneered the use of steam-driven propellers in 1837. By 1841 the first propeller-driven vessel, the *Vandalia*, appeared on Lake Erie and Lake Ontario. Though the *Vandalia* often relied on sails, the propeller engine proved to be more compact and easier to maintain than those needed for rear- or side-paddle steamers. The first propeller steamers on the Great Lakes took advantage of their smaller engines and narrower hulls, which permitted them to pass the Welland Canal between Lake Erie and Lake Ontario. Some propeller advocates predicted that the connection to Lake Ontario and the St. Lawrence River would draw business away from Buffalo and the Erie Canal; this did occur, but not to the extent that some Lake Ontario ports had hoped.[1]

By 1852, the number of propeller steamers on the Great Lakes rivalled the number of paddle steamers. On June 15, near the start of the shipping season, the *Buffalo Daily Courier* announced:

THE NEW PROPELLER—"CITY OF OSWEGO"—This staunch and beautiful vessel will leave this port on her first trip this evening, for Detroit and intermediate ports. She has been constructed during the past winter and spring by Messrs. F. N. & B. B. Jones, for Messrs. E. C. & D. C. Bancroft of Oswego. Her dimensions are, length 143 feet, beam 24 1/2 feet, and her capacity is 3,540 bbls [barrels]. Her engine was built by Barton & Truman of this city. It has a 22 inch cylinder and 3 1/4 foot stroke, with a return flue boiler. The CITY is very neatly finished, being supplied with all the latest improvements in steam vessels. The staining and glazing was done by Mr. C. J. Thurston, and her furniture is from the establishment of Armstrong & Logan, Commercial Street.

She is commanded by Captain William Williams, formerly on the "SYRACUSE," a gentleman and every inch a sailor. This propeller is to be one of a line of twelve propellers, forming Crawford & Co.'s "Northern Railroad Line," running from Ogdensburg to Chicago, touching at Cleveland and Detroit. We doubt whether a prettier or stronger craft of her kind, or a faster one, was ever turned out of our shipyards, or one likelier to add to the reputation of our builders and mechanics, and be a credit to our lake Marine.[2]

Four weeks after launching, on July 9, 1852, the *City of Oswego* left its home port of Oswego, New York on Lake Ontario, bound for Detroit with stops at intermediate ports. On board was a twenty-six-year-old Oswego resident named John B. Green. Green, along with other family members, brought with him his life savings and belongings in order to resettle in the western territories. The only background known about the Green family prior to 1850 comes from accounts written by John B. Green (or, more likely, ghostwritten by another hand based on interviews with Green). In these autobiographical anecdotes, Green made only a few references to his relatives.[3]

His parents, both born in France in the early 1790s, had come to Quebec in the early nineteenth century and settled on a small farm outside Montreal (perhaps Lachine, Quebec, a village on the St. Lawrence, which was mentioned years later by an in-law). Documentation on the Green family in Quebec has not yet surfaced, probably because they Anglicized their given names (and perhaps even the family surname) after they migrated to the United States. According to John Green, his family moved from Quebec upriver to Ogdensburg, New York, in 1835, and five years later relocated to Lake Ontario and the port of Oswego, New York. Green's father made his living there as the operator of a small grocery store.

The 1850 U.S. federal census provides the earliest concrete documentation on the Green family. Thanks to the intolerance of one of the federal census marshals working in Oswego, there is a record of the actual dialogue of the 1850 census interview of the Green household. As an expression of this marshal's ridicule of non–English speakers, the following anecdote can be found in the August 7, 1850, edition of the *Oneida Morning Herald*:

> "The marshal relates the results of his inquiries in a French family as follows:
>
> 'How many persons in this family?'
>
> '*Tree.*'
>
> 'What are their names?'
>
> '*Peter Green, John Green, Lucy Green, Mary Green, Catherine Green.*'
>
> 'But that makes five?—How many in the family?
>
> '*Tree.*'
>
> 'What is your father's name?'
>
> '*Peter Green.*'
>
> 'What is your mother's name?'
>
> '*She is Peter Green, too.*'
>
> No further information could be obtained."[4]

In the 1850 census, the occupation listed for both Andrew and John Green is "sailor," and since the census was taken in July, both sons were probably away from Oswego on a ship or canal boat. Subsequent census records confirm Lucy and Mary as daughters. However, this newspaper clipping indicates the mother was Catherine, while the census document recorded the mother as Eliza. Moreover, later census records from 1860 have the mother's name as Mary, so Mrs. Green's name is not clear, or Peter Green may have had more than one spouse.[5]

All of the Green's family movements placed young John Green close to water, whether it was the St. Lawrence River, Lake Ontario, or the Oswego River and its harbor. According to his own account, he learned to swim in the St. Lawrence at Ogdensburg. It was after the family moved to Oswego that John discovered he had a special talent:

> In the summer of 1841, happening one day to be going along the docks, I saw two men diving for a clock and two boxes of soap that had been stolen and thrown into the river. Thinking it fine sport, I concluded to try my luck. On my first trial I went down about fifteen feet, or half way, when getting frightened, I came up, but seeing the men go down to the bottom, I resolved not to be outdone, I plunged in and reached the same depth, and lo! As my first treasure, I brought up a box of (soft) soap. After diving a few more times, I found where the clock lay, and going again I made a rope fast to it so that it could be drawn up.[6]

Green spent the summers of his teen years diving in Oswego harbor for lost freight—and occasionally for the bodies of unfortunate mariners and canal boat handlers. Green's brief references to those experiences make it clear that at a young age he encountered the gruesome sight of drowned humans and overcame the revulsion of handling bodies. He also saw the comfort that retrieval of those corpses—no matter what their state of decomposition—brought to distraught relatives. On one occasion, Green used his diving ability to save a life: he came to the rescue when a young woman fell off a steamer gangplank while boarding. She sank immediately, and Green had to descend four times before he brought her to the surface, unconscious. Once she was resuscitated, her grateful father gifted Green with $500.[7]

In 1847, Green took the opportunity to run his own canal boat, the freight scow *Illinois*. Green operated it on the Oswego Canal through to the Erie Canal and down the Hudson to New York City.[8] During those journeys, his water skills were occasionally called upon, once to save one of his crewmen who had fallen in the Hudson and once to recover bodies from a sloop that sank in thirty-four feet of water. He also spent five days free diving to recover sixty tons of bar iron lost near Hudson, New York. These three recollections are the only mentions that Green made of his activities during the five years between 1847 and 1852, when he was between twenty-one and twenty-six years old. His written account concludes that period by indicating that he had saved enough during those years to migrate to the American West and go into farming.[9] And so he found himself, with other family members, on the *City of Oswego* on July 9, 1852.

The fate of the *City of Oswego* is at the heart of the history of diving in Lake Erie. What is known about its last voyage comes from contemporary newspapers articles and from the account given by John B. Green. Although there are several newspaper stories about the *City of Oswego*, during that period newspaper editors freely copied from dispatches published in other papers, sometimes with attribution and sometimes without. Any event involving shipping on Lake Erie was likely to be more accurately reported by Lake Erie port papers than by papers of inland cities. To complicate matters further, John B. Green had two different versions of his life story published less than a year apart—a short version and an expanded version—and those two accounts contain puzzling discrepancies about the events surrounding July 1852.

Certain facts are not in doubt: the *City of Oswego* passed through the Welland Canal and reached Lake Erie the morning of July 11. The propeller experienced engine trouble that morning and the ship raised sail once on the lake. It took all day and most of the evening to head southwest toward Cleveland, passing the mouth of the Chagrin River in Ohio about midnight. In the shorter, earlier version of John Green's autobiography, he recalls what then occurred:"We were all awoke about midnight by running in collision with the steamer *America* bound to Buffalo. The propeller went down five minutes after, in forty-three feet of water, drowning thirteen persons, among whom were three of our party. The remaining passengers hardly had time to get on board of the *America* in their night clothes. I dove here three times for the lost ones, but could only find one man and he being so fast to the railing on the promenade deck that I could not release him."[10]

Captain William Williams of the *City of Oswego* was able to launch two small boats to retrieve passengers who had jumped in the water. The *America* took those people aboard and her crew helped to pull other survivors out of the water. After saving as many as they could, the *America* took the survivors to Cleveland. At a later inquest, Williams's crew swore that their lights were visible, that it was a clear, starlit night with calm waters, and that they saw the *America* coming toward them from four or five miles away. As the ships closed, the *City of Oswego* had rung their bell and shouted hails at the incoming steamer. For their part, the crew of the *America* said that they saw no lights and heard nothing. No fault could be determined between these two conflicting testimonies, but there was never a question that human error alone caused the collision.

Captain Williams reported that the passenger list was lost, so the exact number of passengers aboard is not known. The propeller's main traffic that trip was freight, not travelers, so a good guess is that there were only twenty-five or thirty people aboard. John Green's total count of victims matches the count of names listed in the newspaper clippings relating to the wreck. However, Green's account of the number of his family members differs slightly in his two published versions. In the earlier version, he mentions that he was "in company with my brother, his wife and family, making a party of eight." In that version, as quoted above, he bemoans the loss of life, "among whom were three of our party."[11] Green's second version, published less than a year after the first, differs: "I bade farewell to friends and foes, and, with my brother, his wife, and his wife's family, making in all ten persons, left Oswego." In this retelling, "the boat went down in less than five minutes with four of our little party, and many others on board."[12]

Though the complete passenger list was lost, the names of the thirteen dead were published. Curiously, among the dead there was no group of three or four that had family connections to anyone in Green's family. However, even though Green's autobiographies did not list the names of any family or friends that were lost, the victim list did include the names of Ann Green, twenty-two, formerly of Oswego, and her three-year-old daughter, Adelia—but no newspaper mentioned the name of John Green's brother. Instead, several accounts described Ann Green as "Mrs. John Green."[13] Evaluating the evidence, two differing conclusions can be reached. One possibility is that the newspapers all copied each other based on one incorrect source, and that Ann Green was his brother's wife, not his. Since the spelling of names of the victims varied from account to account—indicating they were not copied from one source—this theory's odds of being true must be considered remote.

The simpler explanation is more likely to be true, but shocking: the soon-to-be-famous submarine diver, John B. Green, could not admit that he was unable to save his wife and child from drowning. Ann Green, attired in the era's typical voluminous women's clothing, would have struggled even if she knew how to swim; in addition, she must have been carrying her daughter. No one could have blamed John Green for their loss; whether he blamed himself may be judged from his subsequent actions.

John Green and other survivors returned to Willowick, Ohio, from Cleveland the next day, July 13. Green found his luggage, but it had already been plundered of anything of value. When shipwrecks occurred on Lake Erie, the first known professional salvager on the scene was declared the "wreckmaster." Wreckmasters were charged with recovering everything from the wreck and with preventing pillage. However, these professionals were sometimes beaten to the scene of a disaster by local scavengers. A Cleveland paper cited Green's total property loss as $300; an Oswego paper used the figure $3,000; Green himself, years later, recorded that he had lost $7,000.[14] By that later junction, there was no point in inflating the loss other than to elicit greater pity, but perhaps Green's higher estimation included both cash and lost possessions, such as tools and clothing.

Only a few of the thirteen victims of the *City of Oswego* wreck were recovered the morning of the sinking. The remaining bodies were swept away from the site, drifting submerged. John Green knew from experience the course nature would take on bodies in water. In his first published account, he merely states, "I then watched the lake shore about the scene of the disaster, until the bodies washed ashore, and were sent to their relatives, or decently interred." At no point does he mention anyone accompanying him on this gruesome vigil—one would expect his brother to be there, since he supposedly lost his in-laws. In Green's second, expanded published account, at this juncture in his narrative he goes into a grisly digression on the decomposition of drowned human bodies:

> Although the human body sinks readily when the breath is first exhausted, it assumes another position after it begins to decompose into its elements. The gases which it contains gradually expand the body, and becoming lighter than the water, it begins to rise. The limbs and especially the legs do not expand as much in proportion as the trunk, and therefore inclines the body in the water, until it assumes an almost upright position. It is a sight such as timid souls would quake to look upon—to see a corpse standing

upright beneath the water's surface, its slimy visage swollen, glassy eyes, and rocking to and fro, by every tiny wave that moved the water.[15]

If one considers the strong possibility that John B. Green walked back and forth for miles along the Lake Erie shoreline, day after day, searching for the bodies of his own wife and child, it is hard to fathom how he could deflect the emotion of that experience in such a cold-hearted manner. Perhaps Green was encouraged by a transcriber looking to add melodramatic horror to his narrative, unaware of the truth that Green was hiding. One has to wonder how many times John Green had nightmares about the night the *City of Oswego* sank, and to what degree that event shaped his later career.

Chapter Six

≈

Without Armor and
With Armor (July 1852)

During the week following the wreck of the *City of Oswego*, John B. Green made free dives at the site, working from a small skiff. On July 18, while in the midst of the horrible aftermath, he saw a party of wreckers head out to the site of the wreck of the *G. P. Griffith*, the palace steamer built by Daniel R. Stebbins that had burned down to the water two years earlier. The hulk of the *G. P. Griffith* lay about ten miles from the resting place of the *City of Oswego*, and was just as distant from the shoreline, but at a much shallower depth of only about fifteen feet. Driven by curiosity, Green followed the wrecking crew out to the *G. P. Griffith* and observed as a diver in submarine armor descended into the shallows and stayed under for twenty or thirty minutes at a time. Green asked the crew if he could try the equipment, but was scoffed at by those on the wrecking boat.

Green repeatedly offered to don the armor, and was told that it was so hazardous that he would be risking his life. That was the least convincing counsel Green needed at that moment, considering that he had just lost everything. He told them, "My sins upon my head," and dove into the water without any apparatus, staying down at the wreck of the *Griffith* for nearly three minutes.[1] Green's free-diving skills had no bearing on using submarine armor, but did prove to the others that he was fearless. It is doubtful that any of the wreckers would have allowed Green to try the armor without the consent of the captain of their expedition. If Green had explained that he had just survived a shipwreck but had lost everything, he might have gained a sympathetic ear from the wrecker captain, who was none other than Daniel R. Stebbins, builder and engineer of the wreck they were anchored over.[2]

Ostensibly, Stebbins had let it be known that he intended to try to raise the hull and recover some of the ship's freight. However, given the fact that he was haunted by the memory of that horrible event, and by rumors that his experimental engine lubricant started the inferno, Stebbins was more probably interested in finding evidence of the cause of the fire. In John B. Green, Stebbins might have recognized a fellow damned soul. This might explain why Green was allowed to try out the submarine armor.

Stebbins had contracted with divers, who had brought their equipment to Cleveland. The exact identity of those divers is not known, but it is probably more than a coincidence that Samuel and Eber Ward, owners of Lake Erie's largest steamer line, had recently brought divers to Cleveland to help raise the steamer *Caspian*, which had been dashed against the docks by a storm on July 1, 1852. These divers were from Philadelphia, and brought with them submarine armor made by Howard & Ash Co., using designs specified by John E. Gowen.[3] Gowen had recently used armor of the same design at Gibraltar on the USS *Missouri*. It may be that Gowen had approached the Wards to get a foothold in the lucrative Great Lakes wrecking business.

Green was helped into the armor and descended into the water. He immediately felt anxiety and queasiness, but forced himself to stay down on the wreck for thirty minutes, more to prove himself equal to the efforts of the "old divers" than anything else. On surfacing, Green vomited his breakfast. After resting a couple of hours, he dove again, and this time stayed under for seven hours, far longer than any of the veteran divers among the crew had ever dared. From that point forward, Green became a member of the diving expedition. They stayed at the *G.*

P. Griffith for two weeks and were able to raise some of its machinery and a few coins, but little else of value.

According to Green, he convinced the party to move down the lake to the Silver Creek area, site of Lake Erie's first great steamer catastrophe, the wreck of the *Erie*. The so-called veteran divers among the crew were opposed, once they learned that the *Erie* rested more than sixty feet under the waves. Green related that the other divers thought that their pumps would not be able to maintain air pressure within the armor needed to counter the water pressure at that depth. But Green, based on just two weeks of diving, believed that he could plumb the depths to the *Erie*, or even deeper, without problem. He volunteered not only to be the first to try the dive, but also to do the majority of the underwater work while they were there.

Green was right about the equipment being adequate for that depth. As long as the men cranking the pumps did their job and his air intake hose did not collapse, his armor was able to maintain an internal pressure equal to the water pressure. Once Green commenced work far below on the wreck of the *Erie*, prying once-molten hunks of metal from the scorched timbers and searching the nearby lake bed for personal valuables, doubts began to form in the minds of several of the crew members on the schooner above. Green described what occurred next, and it reads like a morality play on the corrosive effects of greed:

> I sent to the surface large sums of money, which strange to record, invariably proved to me less than when under water. I called the attention of the company to this discrepancy, but was politely informed that it was "very deceiving down there, you know." Perglich, a land shark belonging to our company, and the very one who made the above speech, came to me privately, and saying I "was about right on that point," and if I would agree to hide a part of the treasure below, he would come with me some night, and help secure the same, and we would divide the booty.[4]

Green tentatively agreed to the conspiracy, but was upset a few days later to overhear his partner "Perglich" secretively approach a third crewman, proposing that the two of them would raid the hidden stash before Green was able to return to it. Disgusted, Green says that he opted out of the secret deal and on future dives hid some of the spoils in a lake-bed spot known only to him. While working the *Erie* wreck, Green later related with no sense of shame, he stashed away many franc coins and an enormous slab of molten metal, which he believed to be gold.

It would be fascinating to hear the other sides to this story, particularly from the villain "Perglich." Since Green was unknown to all the others, did the crew on the wrecking schooner suspect Green was holding out on them from the very start? Did they tell Green the amounts recovered were less, in order to steal back a share of what they believed Green had already stashed? Did "Perglich" approach Green on his own initiative to run a separate scam (as Green suggests), or did "Perglich" use this ploy on behalf of all the others to test Green's trustworthiness? If Green was perfectly honest, wouldn't he have exposed the treachery of "Perglich" to the ship's captain and the others?

To complicate matters, Green's first published account of this episode did not assign a name to the duplicitous "land shark." As it happens, the surname "Perglich" has no obvious variant spellings, and as an exact spelling did not belong to any person living in North America in the 1850s. What is known is that there was a man among that crew who would go on to be Green's partner in diving operations for the next year. That man's name was Martin Quigley. After 1853, Green had a falling out with Quigley, and their relationship remained deeply bitter throughout their lifetimes.

Nowhere in Green's memoirs does he mention the name "Quigley," though Martin Quigley played a large role in Green's early diving career. Perhaps Green could not bring himself to mention Quigley by his right name. One possibility is that Green's autobiography was first transcribed in handwritten script, and the editor misinterpreted the scrawl "Quigley" as "Perglich." Alternatively, there is a chance that an editor intentionally changed the name, since the accusation Green was making could be seen as libelous. Finally, if Quigley had been mentioned by his correct name in this one instance, it would have been awkward for Green to omit that he had later partnered in other diving ventures with the same man.

Martin Quigley hailed from Chautauqua County, New York, which borders Lake Erie north of the state line with Pennsylvania. His family had emigrated from Ireland and were early settlers of Chautauqua County. Martin Quigley was born in 1811, fifteen years earlier than John B. Green, and was already middle-aged by the time he met Green in 1852. In the 1850 census, Quigley was residing in Portland, Chautauqua County, about a mile from the Lake Erie shoreline. He married a woman his own age, Lucy Barnes, the product of another family of early settlers of Chautauqua County. The census listed Quigley's trade as carpenter and joiner, a skill that might have led to occasional jobs with shipbuilders.[5] Nothing in Quigley's known history, before or after 1852, would support the characterization of him as a scoundrel.

While Quigley, Green, and the others of the diving party were scouring the wreck of the *Erie* in the summer of 1852, roughly thirty miles away the third of Lake Erie's great nautical disasters of the nineteenth century was unfolding. The burning of the *Erie* in 1841 was the first, the *G. P. Griffith* conflagration in 1850 was the second, and the last was the sinking of the palace steamer *Atlantic* on August 20, 1852. The *Atlantic*, owned by Eber and Samuel Ward, had been added to their steamship line in 1849 to serve the Buffalo-to-Detroit route. The *Atlantic* was far larger than the *Erie* or the *G. P. Griffith*. It measured 267 feet in length, and the breadth of its beam was 33 feet. It weighed, minus freight, 1,156 tons. The *Buffalo Commercial Advertiser* praised it: "We think it far exceeds anything upon the lakes, and affords good evidence of the perfection to which this brand of mechanical business has been brought in our city."[6]

It was not unusual for the Detroit-bound steamer to leave Buffalo with more than 500 persons on board. On the evening of August 19, 1852, it departed with about 450 passengers and a crew of 51.

Nearly 300 of the passengers were immigrants—mainly Norwegians and Irish—headed west to start a new life. As the side-paddle steamer left Buffalo at 9:00 p.m., the weather was calm and the waters were smooth. However, as the evening progressed, fog developed over the lake. At 2:00 a.m., the *Atlantic*, heading southwest, passed Long Point, Ontario, a peninsula that reached out into Lake Erie.

In addition to its precious human cargo, the *Atlantic* carried the life savings of many of the prospective settlers. Some passengers might have deposited their money with the purser, but most keep it in their own possession. The stateroom of the agent of the American Express Company also contained a safe containing over $30,000 in cash and bank notes. Moreover, there were rumors that the luxurious parlor regularly hosted high-stakes card games. The ship itself was financially unprotected; the Wards did not believe that insurance was worthwhile, given how quickly steamboat machinery became obsolete.

A propeller-driven cargo ship, the *Ogdensburg*, was following a northeast course that evening from Cleveland to the Welland Canal. The wheelsman onboard the *Atlantic* saw the lights of the *Ogdensburg* emerge from the fog, but believed his fast ship would pass ahead of the slow freighter. That was a miscalculation; the *Ogdensburg* collided with the port side of the *Atlantic*, gashing a hole. As the two ships struggled to disengage, water poured in through the hole in the *Atlantic's* hull. Captain J. B. Petty ordered the wheelsman to head for Long Point, but the vessel was already sinking. The first-class and second-class passengers had life preservers and floating stools, but the hundreds in steerage-class had nothing. Throngs rushed to the side and jumped overboard; many of them quickly disappeared. Captain Petty and several crew members jumped into two lifeboats and the *Atlantic* began its plunge to the deepest part of Lake Erie.

Because the manifest of the *Atlantic* was lost with the ship, the exact loss of life is not known, but estimates are that between 200 and 300 people drowned. Captain Petty and his crew were assigned most of the blame for their failure to slow when the *Ogdensburg* was sighted, and for their less-than-courageous efforts to carry out their responsibilities as the ship sank. The *Ogdensburg* was not mortally damaged by the collision, and turned around to pull fewer than 200 people aboard. Below them, Lake Erie's richest, most inaccessible shipwreck settled into the dark depths.

Chapter Seven

≈

Mr. Wells's Safe (August–October 1852)

A little over two weeks after the *Atlantic* sank, the unrecovered and uninsured ship was put up for public auction at the Merchants' Exchange in Buffalo. The minimum bid was set at $10,000—a bargain considering the machinery, cash, and furnishings on board. However, the ship lay at a depth of 160 feet, a position so daunting that no professional wrecker deemed it worth the risk and expense. In fact, no one believed it was technically possible to reach the wreck. There were no bids for the *Atlantic*; it remained the property of Eber and Samuel Ward.[1] The Wards announced that they would offer $15,000 to anyone who could raise the vessel.

The Wards almost immediately received a letter from New York City proposing to raise the *Atlantic* with the assistance of a new type of diving bell. The offer came from Henry B. Sears, who was seeking the opportunity to demonstrate the Nautilus diving bell developed in conjunction with the late Edgar W. Foreman. The Wards responded with an invitation to Sears to bring his diving bell to assist in the survey of the *Atlantic* wreck site. Sears quickly arrived in Buffalo in the last days of August, 1852.[2]

In contrast to the Wards, one man was interested in recovering not the whole ship itself, but just one object within it: a safe located in a stateroom behind the *Atlantic*'s wheelhouse. That man was Henry

Wells, founder and president of the American Express Company, owners of the safe. Earlier in 1852, Wells had also founded Wells, Fargo, and Company to offer express services extending west to California. Wells had a pattern of expanding the express business into new territory. In the period around 1840, Wells had worked for Harnden and Co.'s Express and urged them to extend their delivery lines from Albany to Buffalo. The Harndens hesitated, but urged Wells to go ahead on his own, and the result was the American Express Company.

Wells had known and worked with the Harndens at the time Adolph Harnden died in the shipwreck of the *Lexington*. The *Lexington* went down with a cache of Harnden money, a significant loss to the business. Wells was undoubtedly aware that, over the years, different divers, including George W. Taylor and the team of John E. Gowen, had failed to recover the cash from the *Lexington*. Wells was determined that the *Atlantic* would not be his *Lexington*. He wanted his safe from the *Atlantic* and offered $5,000 for its recovery, on top of the Wards' offer of $15,000 for raising the wreck.[3] Wells was also aware of one marine engineer that might be able to accomplish the task.

Wells did not have to travel far from his base in New York to locate the nation's most famous marine engineer, Benjamin Maillefert. In 1852, Maillefert was in the midst of his blasting efforts on the various navigation hazards collectively known as Hell Gate. His efforts there met with great success until March 26, 1852. On that date, Maillefert and four assistants, including his brother, took two boats out to blast a rock hazard known as the "Frying Pan." Maillefert and his brother were in one boat, a patent Francis Metallic Life-Boat, with Maillefert's battery detonation devices. Three other men had the kegs of black powder charges in the other boat, a wooden skiff. Each charge keg held one hundred pounds of gunpowder and sand to weigh it down.

Maillefert's assistants lowered one keg, but it bobbed back up to the surface because it lacked sufficient ballast. They tied that floating keg to their boat and lowered another. The second charge remained submerged in proper position. The workers then handed the detonation wires over to Maillefert; then both boats retreated together away from where the charge had been placed. Maillerfert touched the wires to the battery, but he had been handed the wrong wires. The keg floating beside both boats exploded. The wooden boat shattered and the two men closest to the blast were blown to pieces. A third man in that boat was mortally wounded by flying splinters. In the metal lifeboat, Maillefert and his brother were thrown fifty feet into the air, but were shielded from the flying shrapnel by the metal hull. Both slowly struggled to the surface of the river and swam over to their boat, which stayed afloat thanks to its built-in air chambers.[4]

Maillefert's arm was hurt, and the accident cost the lives of three men. There were some calls for his efforts to be halted; it was also apparent that some in the city—mainly the East River pilots who made a living guiding craft through Hell Gate—had opposed Maillefert's project from the beginning. Despite the criticism, within a month the blasting work resumed, and Maillefert returned to oversee the operations. The same merchants that had hired him rallied to his support, and by July several of the hazards he had blasted were pronounced navigable. When Henry Wells called upon Maillefert's aide in late August to supervise a survey of the *Atlantic* wreck, Maillefert did not hesitate. He reached Buffalo at the same time as Henry B. Sears, just days after the disaster.

Joining Henry Wells, Benjamin Maillefert, and Henry Sears on the survey trip was the new submarine armor diving phenomenon, John B. Green and his diving tenders, Martin Quigley and Charles O. Gardner. Green's exploits at the sites of the *G. P. Griffith* and *Erie* wrecks had not yet been publicized in any newspapers, but word of mouth among the mariners at Cleveland and Buffalo probably brought him to the attention of Eber and Samuel Ward, who recommended him to Wells. The duration of Green's dives and the descents to the *Erie* at over sixty feet were unprecedented operations on the Great Lakes, but may have been less impressive to Maillefert and Sears. Completing the party were two unnamed divers "from New York."[5] There is a strong possibility that these were Charles B. Pratt and an assistant; Pratt had spent the past two years in New York diving over the wreck of the HMS *Hussar* near Hell Gate in search of a treasure trove, so he was undoubtedly well known to Maillefert. And Pratt at that time may have had more years of experience diving than anyone else in America.

Green mentions that these divers brought with them submarine armor that was superior to that which he had used earlier that summer.[6] This is interesting, because all indications are that Green had earlier used new armor made for Thomas Wells and John E. Gowen. If the New York divers brought something superior, then they may have had George W. Taylor's last armor design, which had been used by Whipple and Robinson to good reviews during their Gibraltar survey.[7]

James Whipple, residing in Boston, was a distant but not an uninterested observer of the *Atlantic* salvage efforts. Whipple had a longtime friend, Maine merchant Ephraim B. Grant, write a letter to Eber Ward recommending Whipple as a diver.[8] However, by the time the letter arrived, Henry Wells and Benjamin Maillefert had already brought Sears, Green, and the New York divers to Buffalo. Whipple was, perhaps, not greatly disappointed; he was already planning an expedition to Coche Island off the coast of Venezuela to recover treasure from the Spanish ship *San Pedro de Alcantara*, which sank in 1815. The *San Pedro* had been previously picked over by other treasure hunters with diving bells, but Whipple believed that he could do a more thorough job of searching using a combination of submarine armor and an "ingenious" modified diving bell.[9] The features that made Whipple's bell unique are not documented, but he may have had something like Sears's Nautilus bell.

The *Atlantic* expedition left Buffalo on August 24 on the steamer *Fox* and arrived near where the *Atlantic* had sunk by noon. Green did not enjoy the trip: "While in Buffalo, and on our way to the wreck, these men [the New York divers] were particularly annoying; boasting of what they had done, going down in 150 and 200 feet of water. . . . They termed me 'the fresh water diver'; that I played in shallow water, like a boy who could not swim, dared not venture where it was too deep. Thus I was annoyed by their boasting, at work, at meals, and at rest."[10]

The expedition's first challenge was finding the wreck, since no one had yet marked the site with buoys. They dragged lines over the area for two hours, without success, before breaking off due to bad weather. They resumed the search later in the day, but still had no success. However, by the time the next day's edition of the *Cleveland True Democrat* was published, it was reported that the wreck had been found in 160 feet of water.

Once that sounding had been taken, the New York divers refused to dive to that depth, much to Green's bemusement. Green volunteered to make the descent, an offer that Benjamin Maillefert must have considered carefully. Maillefert had lost three men earlier that year; and Green was proposing to conduct a dive in armor he had never used before—a dive that more experienced men declined to risk. Green must have been convincing, for Maillefert let him proceed. Green was lowered to 105 feet, deeper than he had ever gone before, but his air hose did not work properly. Over the next several days, repairs to the hose and bad weather prevented further dives.[11]

On Saturday, September 18, Green descended again and his lower body dropped directly into one of *Atlantic*'s upright smokestacks. Immobilized and held above the top of the pipe by his armpits, he signaled to be lifted. After being laterally repositioned, he dove again and this time reached the *Atlantic*'s braces, but the ship above, the *Fox*, from which Green was suspended, was rocking back and forth. The motion transferred down to Green, creating a danger that his hoses and tether line might get tangled in the wreckage. After another interval of waiting on the *Fox*, Green went down yet again and reached the deck at 152 feet. At that depth there was no visibility; Maillefert had brought a prototype of an underwater lamp, but it proved ineffective.[12] The *Buffalo Commercial Advertiser* claimed that Green's descent was the deepest dive ever made, the previous record having been 126 feet.[13] The assertion is difficult to prove, but there is no doubt that Green went as deep as indicated. However, at that point Green's makeshift air hose burst, and he had to be immediately brought to the surface.

There is no documented evidence that Henry B. Sears tried to descend to the *Atlantic* in his Nautilus diving bell. If he had made such an attempt and met with even mixed success, it probably would have been noted. The bell he brought had only been built and tentatively tested in the past two months, so it was still considered experimental. The likelihood is that Sears was unprepared to go to that depth, out of caution or lack of adequate hose length. That left Maillefert's efforts on behalf of Henry Wells entirely on the shoulders of John B. Green.

The survey party returned to Buffalo, where a stronger air hose for the submarine armor was constructed. Maillefert also wanted to use a larger ship, to provide Green with a more stable diving platform. Green talked to newspapers for the first time, and described the sensation of diving to such great depth: "I found great difficulty in moving; the water was so compressed; and with the diminutive air-pipe which we used, it was next to impossible to keep the armor inflated below the waist, and often it rose as high as the chest. The pressure was immense. The rush of blood to the head caused sparks of various hues to flash before my eyes, and I had a constant tendency to fall asleep, although the pressure on my limbs was enough to crush them under ordinary circumstances."[14]

Wells, Maillefert, and Green left Buffalo on September 30 aboard a different ship, the *Columbia*, and returned to the wreck site. Green made eight dives, and reached the upper deck of the *Atlantic*, but was unable to determine his precise location. However, the physical stress made it impossible to stay more than a few minutes at a time, and he could not make any further progress. The expeditions had already cost between $3,000–$4,000, and Henry Wells was discouraged that a diver in submarine armor would ever be able to succeed. Green's attempts were halted and the party returned to Buffalo. Henry Wells and Benjamin Maillefert were now convinced that the only way to recover the safe would be to raise the *Atlantic* itself.

In November of 1852, Henry Wells and the American Express Company contracted with Albert D. Bishop, inventor of Bishop's derrick, to raise the *Atlantic* and bring her to a dry dock within three months of navigation opening on Lake Erie in the 1853 season. The price the parties agreed to was $25,000.[15] Both Wells and Bishop were aware that, for the derrick to lift anything, chains would need to be attached to *Atlantic's* hull; and that work could only be accomplished by submarine armor divers or by a diving bell. Bishop was betting that affixing the chains would be an easier task than asking a diver to find the bursar's office and retrieve the safe. Wells, for his part, was simply relieved to contract out the work, since he had already lost too much money trying to get the job done himself. Bishop had the winter of 1852–53 to construct the largest version of his crane he had ever attempted—he would need a gigantic derrick to lift the *Atlantic* from its grave.

Chapter Eight

~

The *Erie* Jinx (1853)

Daniel R. Stebbins, builder and engineer of the ill-fated *G. P. Griffith*, sought to explain the cause of the fire that had destroyed his ship. He led a party of divers to the wreck in July of 1852, and there introduced John B. Green to submarine armor diving. Following that expedition, Stebbins tried to quell the ugly rumors that he had caused the fire with an untested engine oil, but he had trouble getting his story out. After John B. Green's name and reputation became famous as a result of the dives to the *Atlantic*, Stebbins asked Green to take the results of what they had found at the *G. P. Griffith* to the press. In December, 1852, several papers related the identical story after interviewing Green: "The fire, now that the wreck has been examined, is supposed to have risen from the ignition of a large quantity of friction matches which were in freight without the knowledge of the persons in charge of the boat, since the wood of the boat remaining unburnt was found to be thoroughly impregnated with sulfur. In sawing the keelson, the friction of the saw several times ignited the wood."[1]

Unfortunately for Stebbins, this explanation did not gain traction, since some questioned why anyone would be transporting matches to a region where matches were manufactured and readily available. Moreover, if all the matches had ignited, they would not have left a residue of sulfur sufficient to reignite wood.[2] Whether Stebbins believed he had been vindicated or not, the horrors of the *G. P. Griffith* never left him. He abandoned all ambition to earn his fortune from plying the waters of Lake Erie and spent his remaining years as a mechanic for grain elevator companies in Toledo.

Green and Martin Quigley visited several newspapers in December, 1852, and January, 1853, and did not limit themselves to the subject of the *G. P. Griffith*. They announced that they would be working with Albert D. Bishop the coming season in his efforts to raise the *Atlantic*, and that an attempt would also be made to put tackle on and raise the *Erie*. They brought with them some of the coins found on the *G. P. Griffith* and the *Erie* that they had found the previous summer. Green's diving endurance and the depths he reached were also mentioned in these articles.[3] In some cases, editors concluded their pieces by anointing Green with almost supernatural skill. The *Buffalo Commercial Advertiser* said "He seems calculated by nature for a Diver, his powers of endurance being truly wonderful."[4] According to the *Cleveland Herald*, "John Green alone, in all the world, possesses the *secret* power to 'go down to the depths of the seas.' . . . he is confident that he can reach the depths of any of the Lakes at their greatest soundings, explore vessels and attach to them the necessary fixtures for raising."[5]

John Green had been boarding in Cleveland ever since the sinking of the *City of Oswego*, except during the lengthy times when he was working from ships on the lake. His first dive of the 1853 season took place at the Cleveland docks in April, before regular shipping had opened on the lake. Green assisted Captain William Nelson, a dockmaster employed by the Wards, in setting charges on the wreck of the *Caspian*. Removal of the *Caspian* had defied the efforts made by divers brought to Cleveland the previous summer, and it was still creating a navigation hazard to other ships coming in to port.[6]

Once shipping opened for the 1853 season on Lake Erie, Green and Quigley, who now owned their own submarine armor, returned to the *Erie* to retrieve the coins and molten metal pieces that Green had hidden the year before. Green was crestfallen to discover that the huge metal slab he had concealed, which had appeared as shiny gold underwater, when brought to the surface turned out to be the remains of the ship's brass bell (proving, as Green had been told a year earlier, that it was "deceptive down there"). Over several weeks, Green and Quigley sailed from the *Erie* site back and forth to Silver Creek. They had the *Erie* site marked with buoys, both for their own use and to assert an active salvage claim.

However, one day in July they arrived to discover their markers had been cut and that another dive crew was working the wreck. That party included several men whom Green and Quigley had worked with the previous year. Threats were exchanged before an uneasy truce determined that Green and Quigley would work one end of the *Erie* and the other divers could work the opposite end. When a rival diver approached Green underwater, Green thrust his crowbar menacingly at him. The two parties kept their distance after that incident.[7]

After retrieving over $700 in francs (from the *Erie*'s French and Swiss victims), Green and Quigley proceeded to Buffalo to discuss the plans for raising the *Atlantic* with Albert Bishop and to help complete construction on the gargantuan derrick. However, after being there only a short time, Green and Quigley were called back to Silver Creek at the request of the rival divers they had left at the *Erie* site. One of those divers, a man named William McDonnell, had gone down and failed to respond to signals from the surface. His assistants tried to haul him up, but his body was caught on something below. It was days before Green and Quigley were called in to help recover the body—and the expensive submarine armor that McDonnell was wearing.

Arriving at the scene, Green and Quigley found that McDonnell's hose and tender line had been entangled in the marker buoys, but that they had come loose on their own by the action of waves and current. They hauled up the body and found the man's head purple and swollen, with blood forced from the mouth and ears. They brought McDonnell back to Silver Creek, where an inquest determined that he had suffocated from lack of air—either the air pump was inadequate or the hose had failed.[8] However, the swollen head and rush of blood indicate that McDonnell could have experienced some degree of "diver's squeeze," that is, an abrupt pressure differential. In extreme examples, from much deeper depths than McDonnell descended to, diver's squeeze can violently crush the body and force blood and viscera up into the hard helmet.[9]

John Green and Martin Quigley must have returned to Buffalo in a somber mood. The *Atlantic* lay at two-and-a-half times the depth of the *Erie*, and would require dozens of dives to affix the lifting network of sweep chains. The dangers, known and unknown, were far greater than those to which William McDonnell had succumbed. There was no guarantee that Bishop's derrick would be capable of the task, as a Buffalo paper detailed in an article headlined "Something of an Undertaking":

> In the creek [Buffalo River], a short distance above the Clark and Skinner canal may now be seen the Derrick, constructed by Mr. A. D. Bishop, and nearly finished, which is to be used on Monday next in an attempt to raise from the bottom of Lake Erie, what may yet remain of the ill-fated steamer *Atlantic*.
>
> The hulks of two steamers, the *Madison* and *Lexington* [a different steamer than that which sank in Long Island Sound], have been taken for the purpose, thoroughly repaired, and rendered perfectly sea-worthy, these are connected by two bridge trusses across the decks, converging to an acute angle on the *Lexington*, and so diverging that on the *Madison* they are sufficiently wide apart to allow for the machinery to be erected between them, and it is by these trusses that the machinery is supported in an erect position. The Derrick is built up on the *Madison*, and is of several times the capacity of that erected on the opposite shore [a dock derrick at Buffalo], and a little above Fish's elevator.

It weighs sixteen tons, stands eighty-five feet from hoist to the water, the whole height being one hundred and forty-two feet. The lifting machinery is driven by two oscillating engines of eight horsepower each, which by multiplied gearing gives a power equal to an engine of *five hundred and seventy-six horsepower*, and this by multiplication of pulleys, gives the whole machinery, a lifting power of 4,960 tons—This estimate, of course, does not allow anything for friction, which, when overcome, will somewhat lessen the amount of power which can be made available, but all unite in the opinion that there is, after friction is overcome, a vast surplus power over what will be required to raise the vessel.

The hulks of the two steamers are distant from each other seventy-two feet, from centre to centre, between sides, and the bridge trusses are framed in and firmly fastened by means of ponderous iron bolts and bars, to the hulk of the *Madison*, while the *Lexington* is secured to the angle of the trusses by means of what mechanics call a "universal joint," leaving the latter vessel free to turn in any direction or even to make a rolling side movement, and yet be kept in her proper place, and under management of her helm. The object of the *Lexington* is to form a counterpoise to whatever weight may be attached to the Derrick, on the other side of the *Madison*. Coal and iron are now being taken on board the *Lexington* in large quantities for ballast. The *Madison*, it is said, has a measurement of 1,000 tons in addition to the machinery and the weight of the engines now on board. The two boats will be floated down the creek sometime today, and those wishing an opportunity to examine these novel craft and their machinery, previous to their departure for the scene of labor, will thus have an opportunity, which it may be will not soon occur again.

> The steamers *Baltic* and *Hendrick Hudson* will take
> the Derrick in tow early on Monday morning; and the
> kind wishes of thousands will accompany Mr. Bishop to
> the scene of his labors. Though many doubt the success of
> his undertaking, yet there is not one but that heartily wishes
> the effort may be successful.[10]

The *Baltic* left Buffalo on August 22, 1853, with Bishop's derrick in tow, without aide from any other steamer. The wind was blowing when they arrived at the *Atlantic* site, so they anchored near Long Point for the night. At the first dive opportunity, Green attempted to get chains under the bow, but that end of the ship was too deeply embedded in sand. They decided to leave the chains down on the bottom, fastened to ropes leading up to buoys on the surface. Wind kept them idle until August 28, but when they returned to the site they found that their buoys—and presumably their chains—had been swept away. Frustrated, they found they had to return to Buffalo to secure new chains.[11]

Waiting in Buffalo for Albert D. Bishop was a letter from James A. Whipple, whom Green had supplanted as the most famous diver in America. Writing from Boston, Whipple offered his assistance, as he had offered it to the Wards the previous year. Whipple's letter mentioned that he had worked with Bishop's brother recovering the engine mountings of the *Pioneer* lost in the Hudson River in 1851. Whipple's offer was also delivered in person by one of his assistants, Daniel Driscoll. Driscoll later reported to Whipple that he met with both Albert Bishop and American Express's Henry Wells, who had come back to Buffalo to observe Bishop's efforts. Driscoll urged them to call in Whipple to make all future dives; however, Bishop was inclined to give Green another opportunity.[12]

Driscoll also related back to Whipple that Henry Wells and the American Express Company, for their part, wanted no more to do with Green. In a letter, Driscoll pleaded with Whipple to come to Buffalo as soon as possible to take advantage of the situation.[13] A longtime family friend of Whipple's, a shipbuilder from Gardiner, Maine, named Peter G. Bradstreet, also had contacts among the shipbuilders and ship owners in Buffalo. Bradstreet wrote to Whipple that Wells had no confidence in "Green the Frenchman." Despite this encouragement, Whipple hesitated and remained in Boston.

By September 7, Bishop, Green, and the derrick were back working over the *Atlantic*. They made better progress getting the chains around the vessel, but with no other surface ship to assist the *Baltic*, the chains became hopelessly entangled. Bishop called the effort off and had the party retreat to Buffalo. He later told the newspapers he was not yet discouraged, but believed it might be prudent to perfect the chain-laying technique on a more accessible wreck, the *Erie*. Consequently, Green and Quigley were dispatched from Buffalo on September 12 aboard the steamer *Southerner* to return once again to the wreck that was now familiar to them.[14]

Once Green and Quigley confirmed that they were successful in attaching the chains, Bishop set out on September 23 on the steamer *Empire* with the derrick in tow. When the *Empire* arrived at the *Erie* site in the afternoon, a storm blew in from the west. The *Empire* and the derrick were anchored and prepared to ride out the bad weather. However, by 3:00 a.m. on the 24th, the *Empire* was rocking so violently in the gale that it lifted anchor and headed for harbor. Bishop, Green, and several other men stayed on the *Madison*, the steamer supporting the derrick. At 4:00 a.m., the trusses that connected the *Madison* and the *Lexington* broke, and the *Lexington* drifted closer to the other boat. With the counterweight gone, the giant derrick slowly pitched over and then snapped off, the free piece slamming down on the *Lexington*.[15]

Both the *Madison* and the *Lexington* began to sink, along with the remains of the derrick. Bishop, Green, and the remaining crew had time to launch the two lifeboats, but a horse that was on board the *Madison* to power the hoist treadmill went down with the ship. One of the lifeboats was able to sail away to safety, but the other drifted with men clinging to it until they were picked up by the *Empire*.

Bishop was devastated. The loss of the derrick, the two steamers, and the equipment on them amounted to over $30,000.[16] His dreams of becoming the James Eads of the Great Lakes lay in ruins. Bishop was soon beset by creditors, and retreated back to Brooklyn to restore his business. He never ventured to the Great Lakes again. Henry Wells, disgusted with the salvage endeavors, washed his hands of any further attempts to recover the safe on the *Atlantic*. James Whipple canceled any plans he had to travel to Buffalo—but also kept Daniel Driscoll there to report on any further developments.

Green and Quigley were not untouched by the derrick disaster. For reasons unknown, their partnership did not survive past September, 1853. Almost certainly, some of their submarine armor equipment was lost when the derrick sank, and this might have caused a financial dispute between the two men. Quigley, now a skilled diver himself, opted to operate out of Detroit, where a surplus of wrecking jobs awaited. Green, in need of cash, in November of 1853 returned to the site of the *City of Oswego* wreck, where his life had changed the year before. The water was cold and muddy, but he was able to recover $2,500 in cargo to tide him through the winter.[17]

1. "Steamship *Erie*." A superb example of maritime art. [Courtesy National Gallery of Art, Washington.]

THE STEAMBOAT ERIE,

As she appeared while wrapped in flames, on the evening of the 9th August, 1841, at the mouth of Silver Creek, Lake Erie, about thirty miles from Buffalo, when about 200 persons perished.

2. "The Steamboat *Erie*." From an engraving by Huestis & Craft, New York, 1841. The artist likely based this depiction on a copy of the painting of the *Erie* made for her launching, adding flames and figures in distress. [Courtesy New York Historical Society, Maritime History Collection. NYHS Image #44914]

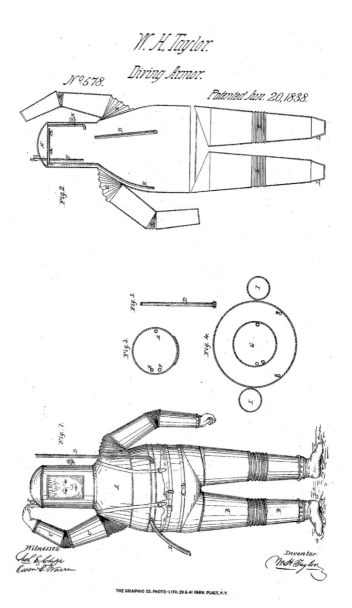

3. William Hannis Taylor's 1837 U.S. patent drawing for submarine armor. The drawing depicts connected rigid metal hoops, but would not have performed as an atmospheric diving suit. Taylor and his partner, George W. Taylor, likely modified their suit in imitation of the English Siebe diving dress. [Courtesy United States Patent and Trademark Office.]

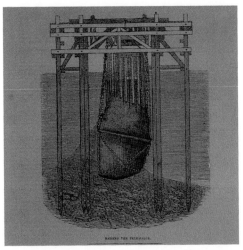

4. "Raising the *Telemaque*." The failure of W. H. Taylor's treasure hunt obscured his clever engineering approach. Tides precluded the use of ships as raising platforms, so Taylor constructed a pier around the wreck. Chains could only be fit around the ends, so to lift the center section of the hull, he drove dozens of chain-fitted harpoons down through the timbers. [*The Illustrated London News*, December 17, 1842, p. 505. From the collection of the Author.]

5. "Submarine blasting with the voltaic battery." Removing obstructions from harbors and rivers, whether they were made of rocks or hulks of ships, was a fundamental marine engineering operation. Benjamin Maillefert mastered the technique of using concussive charges set in the water above obstructions, but in many cases charges would need to be drilled into the bed. [Image reproduced with permission of the American Antiquarian Society.]

6. James Aldrich Whipple (1826–1861). For a brief time, Whipple was the most famous diver in America. For months, Whipple waited in anticipation of taking over the *Erie* and *Atlantic* diving operations from John B. Green. [Photo courtesy of Alice J. Murphy]

7. James B. Eads, from "Sketch of James B. Eads," *Popular Science Monthly*, v 28, Feb 1886, p. 642. [From the collection of the author]

8. "Hell Gate and Its Approaches." Close-up view from United States Coast Survey Office map. [Courtesy NOAA Office of Coast Survey.]

FLOATING DERRICK—RAISING THE WRECK OF THE OLIPHANT, EAST RIVER, NEW YORK.

9. "Floating Derrick," *Scientific American*, Feb 7, 1880, vol 42, no 6. Albert D. Bishop's derrick design remained basically the same from the early 1850s through the 1870s. The above depiction is very similar to Bishop's 1846 patent drawing. For raising the *Atlantic* and the *Erie*, Bishop installed the derrick on two old steamer hulls, braced together, rather than on a single barge as shown above. [From the collection of the author.]

10. "The 'Griffith' Steamer, Passing the Lighthouse at Buffalo Harbour Point," *The Illustrated London News*, Jul 27, 1850, p. 84. The second of Lake Erie's three great wrecks. [From the collection of the author].

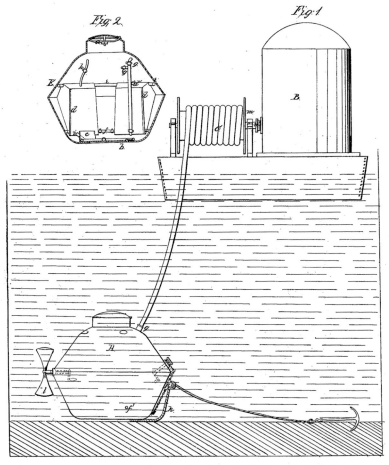

11. The patent drawing for the Nautilus diving bell. Note the bell is not suspended, and has a small hand propeller to move itself within a limited area. Air intake hose is still supplied from a compressed air tank on the support ship. Air chambers in the bell control its buoyancy. [E. W. Foreman, patentee. "Diving Bell." U. S. Patent No 9,965. Patented Aug 23, 1853.]

12. Interior View of the Henry B. Sears Nautilus Diving Bell. From *Frank Leslie's Illustrated Newspaper*, April 2, 1859. The Nautilus represented a hybrid diving bell/submarine. Author Jules Verne was said to have seen the Nautilus diving bell, and used it along with Robert Fulton's Nautilus submarine concept, and early French submarines of De Villeroi and Payerne, as the basis for *Twenty Thousand Leagues Under the Sea*. [From the collection of the author.]

13. Lodner Philips' 1859 Submarine concept, from *The Illustrated London News*, March 12, 1859, p. 272. After interest by the United States Navy was dropped, Phillips attempted to interest British officials. The illustration above represents a concept far larger than Philips' last working model, the *Marine Cigar*, which was lost in Lake Erie in 1853–1855. [From the collection of the author.]

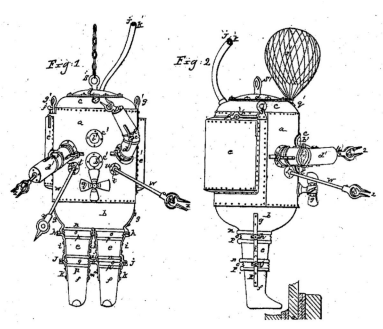

14. Lodner Philips' *Marine Cigar*, as described in *Scientific American*, Feb 12, 1853, volume 8, issue 22, p. 172. This is likely the most accurate depiction of the submarine used on Lake Erie. The length is not described, but visual clues suggest it measures less than thirty feet. [From the collection of the author.]

15. Lodner Philips' 1856 Submarine Armor patent. This rigid design represents a true atmospheric diving suit, rendering the diver free from complications of breathing pressurized air. However, any breach would be catastrophic. [Courtesy United States Patent and Trademark Office.]

Man in **WELLS & GOWEN'S**
Submarine Armor under Water.

16. "Man in Wells & Gowen's Submarine Armor under Water." Etching commissioned by Thomas F. Wells. Note the diver's traditional tools, the pike and rope. It appears that this suit was worn without the familiar duck canvas outer layer. [Image reproduced with permission of the American Antiquarian Society.]

Man in WELLS & GOWEN'S
Submarine Armor, recovering a lost Anchor

17. "Man in Wells & Gowen's Submarine Armor, recovering a lost Anchor."
Etching commissioned by Thomas F. Wells. An air hose and signal line tether the
diver. In this case, the diver descended via a rope ladder. In other cases, a separate
tow line lowered the diver. [Image reproduced with permission of the American
Antiquarian Society.]

18. "Collision between the Steamer *Atlantic* and Propeller *Ogdensburg* on Lake Erie, N.Y." [*Gleason's Pictorial*, September 11, 1852. From the collection of the Author.]

19. John Tope (1822–1854). The horrible death and disfigurement of amateur diver William McDonnell might have been dismissed as lack of experience, but when the same happened to seasoned diver John Tope, John Green must have realized the unforgiving risks he took with each dive. [Image supplied by S. White (Australian descendant of John Tope)]

20. "View of operations, preparatory to raising Wreck of Steamer *Erie* sunk in Lake Erie." Engraving prepared for Thomas F. Wells of Wells & Gowen. One of the divers, it can be assumed, is John B. Green. Note that in place of a derrick, Wells & Gowen used two trusses mounted between ships. [Image reproduced with permission of the American Antiquarian Society.]

21. "Towing Wreck of Steamer *Erie*, sunk in Lake Erie in 1841." Engraving prepared for Thomas F. Wells of Wells & Gowen. In reality, only three-quarters of the hulk made it to Buffalo (and there are no Lake Erie whales). [Image reproduced with permission of the American Antiquarian Society.]

22. Torpedo Station, James River, Va. Prof. Maillefert and naval officers who were employed in removing Confederate torpedoes, April, 1865. [Library of Congress Prints and Photographs Division Washington, D.C. 20540 USA http://hdl.loc.gov/loc.pnp/pp.print. LC-DIG-ppmsca-33153 (digital file from original item) LC-B8184-434 (b&w film copy neg.)]

23. "Underwater work in Charleston Bay to clear entrance to the port." 1870 illustration from a French magazine. Though the depiction is exaggerated, it conveys the herculean task facing Elliot Harrington during the war and Benjamin Maillefert after the war. [Image courtesy of Picture Collection, New York Public Library, Astor, Lennox and Tilden Foundations.]

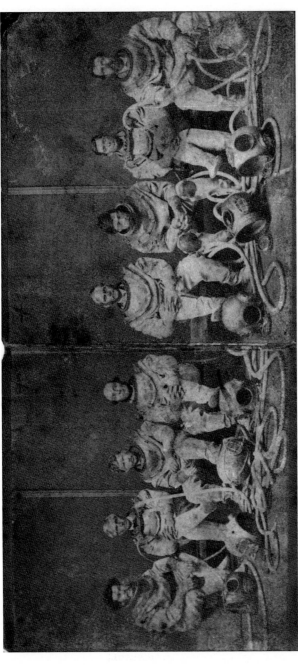

24. Studio portraits of Detroit divers, circa 1875. Preserved as one image, this is actually two separate photographs taken with four sets of diving apparatus. Elliot P. Harrington is fourth from left; his brother-in-law James Phillips is the next figure to the right. The group likely is the Detroit expedition that went to the James River in Virginia in search of the safe of the USS Cumberland. Note the four sets of armor are of different design. [Courtesy Archives & Special Collections, Daniel A. Reed Library, SUNY Fredonia.]

25. Augustus Siebe (English) helmet circa 1850s–1860s. [Courtesy of the Leslie Leaney Collection. Photograph by Robert Noriega, Brooks Institute of Photography. ©2014 Leslie Leaney Archives. All Rights Reserved.]

26. Charles B. Pratt helmet, 1850s–1860s. Manufacturer unknown, the lightly-used helmet appears to copy many Siebe features, but has an unusual arrangement of intake directly over exhaust on the exact rear of the helmet. Possibly made to Pratt's own specifications before he retired from diving in 1871. [Photography by the author courtesy of the Worcester Historical Museum.]

Chapter Nine

≈

Harrington and the Diving Boat (October 1853–Spring 1854)

As evidenced by the clash of salvage crews over the *Erie* wreck during the summer of 1853, John B. Green and Martin Quigley were by no means the only divers on Lake Erie. However, either none of the other divers had much daring or skill, or they were limited by inferior equipment. The unfortunate William McDonnell died on only his second venture underwater, and the day before, another amateur from the same crew had lost consciousness while submerged and had to be hauled up and revived. However, during the same period another Lake Erie diver demonstrated his skill, though his exploits escaped notice by newspapers until years later.

His name was Elliot P. Harrington, and he hailed from Westfield in Chautauqua County, New York. Westfield lay about a mile from Portland, where Green's partners Martin Quigley and Charles O. Gardner resided in the early 1850s.[1] There is no surviving documentation that Harrington worked on diving operations with Quigley or Gardner (or Green) until 1856, but it strains credulity to consider that along the whole length of Lake Erie, with its large cities, three of the earliest divers came from the same small, rural Chautauqua county area without having knowledge of one another.

It is tempting to surmise that Harrington was among the diving crew that John B. Green first encountered while they were working on the *G. P. Griffith* in the summer of 1852. Technically, that might have been possible; but in July of 1852 Elliot Harrington was, by sad coincidence, dealing with a personal catastrophe of his own. Harrington, like Quigley, came from a large family of Chautauqua County settlers. In the years prior to 1852 he operated a dry goods store in Westfield, where he sold tools and also made carriages. On July 14, 1852, one day after the *City of Oswego* sank with John Green's family, Elliot Harrington's Westfield store burned down. Some faraway newspapers, such as those in Boston and New York City, noted both calamities in adjoining paragraphs—an ironic coincidence given the later careers of Green and Harrington.[2] Harrington's business was insured for $1,000, but he lost $1,300 in property.[3] Idled from his vocation by disaster, it is possible that Harrington sought temporary employment on a salvage crew with other men from his town.

By Harrington's own account, his first experience in submarine armor came in November, 1852, in the wake of a momentous Lake Erie gale that wrecked two dozen vessels. The propeller ship *Oneida* was overcome by waves and started to flounder near Dunkirk, New York, with a load of flour. According to one source, Harrington helped recover the cargo, much of which was sealed in watertight barrels.[4] However, over the years, Harrington gave out conflicting information about his first dive. Curiously, when recalling his diving exploits in later years, Harrington skipped mention of any diving activity in 1853 or 1854. However, he did make an elliptical reference to a unique underwater venture that could only have taken place between late October, 1853, and the spring of 1855, immediately following the failure of Bishop's Derrick at the *Atlantic* site and its subsequent sinking over the *Erie*.

Lodner D. Phillips, the shoemaker turned submarine inventor, had completed his third design, the *Marine Cigar* in Michigan City, Indiana, between 1852 and 1853. Phillips was shy of publicity: the testing of his submarines was never generally announced, and he never spoke to newspapers. Anecdotally, his relatives have handed down stories that the *Marine Cigar* was tested successfully in Lake Michigan, and that on one occasion he took his whole family out for a voyage for an entire day. However, Phillips had applied for a patent in 1852 for the vessel's steering mechanism, and therefore had an interest in proving the value of his design.[5]

When Bishop's derrick sank in September, 1853, Phillips saw an opportunity to prove the worth of his submarine. If he could help raise the *Atlantic*, the most challenging underwater operation of the age, his services would be in high demand. The expense of moving the *Marine Cigar* from Michigan City, Indiana, on Lake Michigan, to Lake Erie—as well as room and board while conducting operations there—would be considerable. It is unlikely that Phillips would have embarked on such a momentous undertaking without contacting and getting the consent of Henry Wells or the Wards. If Henry Wells was the man Phillips approached, his willingness to let Phillips attempt to reach the *Atlantic* could be seen as desperation.

Despite Phillips's secretiveness, the transport of the *Marine Cigar* did not escape notice. In mid-October, the *Detroit Advertiser* noted:

A SUB-MARINE PROPELLER—We saw, yesterday, at the Railroad Freight House, a curious looking structure of wood and iron, shaped something like a pear, only about twenty feet long, with a little propelling paddle-wheel at one end, and an iron flanged steering paddle at the other. On the sides were small bull's eyes windows, filled with very thick glass. The machine, we were informed, is Phillips' Sub-Marine Propeller, and came over by the railroad from Michigan City, on its way to pay the *Atlantic* a visit. We know not whether to examine her previous to making new efforts to raise her, or merely to ascertain legally whether she was hit by the *Ogdensburg* on the larboard or starboard side, which is a point on which there is some question yet.[6]

By October 26, the *Marine Cigar* had been shipped to Buffalo, where again a dock observer took note: "We noticed Phillips' Patent Diving Boat on the Michigan Central Dock this morning, having been brought down by the steamer *Ocean*. The owner of the boat intends making an attempt to reach the *Atlantic*, in a day or two with this new invention."[7]

Since the above two articles are the only documented evidence of the *Marine Cigar* on Lake Erie, the outcome of Phillips's submarine venture must rely on surmise and anecdote. One assumption that can be made is that Phillips required the assistance of at least a few Lake Erie mariners to be guided to the *Atlantic* wreck site. Getting to the site from Buffalo probably involved a tow from a support ship. A written statement that diver Elliot Harrington made in May, 1863, suggests that he was one of those who assisted Phillips with the *Marine Cigar*. In a letter to Rear Admiral Samuel Francis Du Pont, in which Harrington offered to build a submarine for the U. S. Navy, Harrington stated, "Among [my] experiences was one with a submerged small propeller, driven by hand power, capable of being supplied with air by means independent of all outside help. With it I can make 1 1/2 miles per hour at a depth of 80 feet or less, and could conduct operations outside of it at any given depth with success."[8] There are no other explanations as to where Harrington might have gained this experience, if not with Lodner Phillips aboard the *Marine Cigar* in Lake Erie. Harrington spent the years 1857 to1863 living in Charles City, Iowa, where (it is safe to say) there were no submarines. In the years prior to 1857, Lodner Phillips was the only American experimenting with a vessel such as the one described, and Harrington was then residing in proximity to the area where Phillips was operating.

The only surety concerning the fate of the *Marine Cigar* is that it did not salvage anything from the *Atlantic*. There are no contemporary accounts of its trials, but many years later, in 1887, one of America's leading submarine experts, Edmund L. Zalinski, related the story he had heard: "Attempting to reach a wreck in water 155 feet deep, he [Phillips] found his boat leaking badly when at a depth of 100 feet, he therefore returned to the surface. He lost his boat in a subsequent test, where he lowered it, attached to a hawser, in attempting to lift it, the hawser broke and the boat remained. Fortunately there were no occupants."[9] Zalinski partnered with John Philip Holland to form America's first submarine company in the 1880s. His career was based in New York City, where Lodner Phillips spent the last decade of his life, so Zalinkski might have heard about Philips from local sources in the city.

While Zalinski could be considered a very reliable source, another anecdote surfaced in the 1890s in the pages of the journal *Marine Review*. A letter was sent to the editor by a former tender, T. E. Kinney, who had assisted Elliot Harrington in the late 1860s. Kinney's letter contains several egregious factual errors, but he does mention something interesting about the efforts over the *Atlantic* wreck: "Then a man from Chicago who had a patent cast iron armor, or diving dress, made several unsuccessful attempts to reach the vessel. The owners of this outfit left the wreck for the purpose of getting additional supplies."[10]

Phillips had been a resident of Chicago in the early 1850s; it was there that he filed his patent applications. Kinney appears to be making reference to Phillips's "Submarine Exploring-Armor," granted U. S. patent 15,898 in 1856 (but applied for much earlier). Kinney's suggestion that Phillips was testing not only his submarine, but a solid atmospheric diving suit, in Lake Erie is amazing—these technologies would not be accepted as practical inventions for decades.[11] However, the *Atlantic* was too steep a challenge for Phillips's prototypes.

Even James Whipple's spy, Daniel Driscoll, who wintered in Buffalo in 1853–1854, reported on the presence of Phillips's submarine. Driscoll wrote on March 2, 1854, "There is a Submarine Boat here from New York."[12] Driscoll was wrong about New York, but he was undoubtedly referring to the *Marine Cigar*. Since Driscoll did not report that he had seen the craft in person, it is impossible to tell if the *Marine Cigar* had already met its final accident or still survived into 1854. At any rate, Driscoll's employer, James Whipple, was not enticed to come to Buffalo to try his hand at reaching the *Atlantic*. He returned to the

Venezuelan coast over the winter and the wreck of the *San Pedro de Alcantara*.[13] Later in 1854, Whipple received a very lucrative government contract to construct the sea wall and dock at the naval station in Pensacola, Florida.[14] In Whipple's mind, the risks of going after the safe on the *Atlantic* made it a bad proposition, especially when so much other underwater work was available and offered better chances of returning a profit. His assessment was not wrong.

Chapter Ten

≈

Boston Bliss (1854–July 1855)

In January of 1854, John B. Green left Cleveland for a trip to Boston.[1] While there, he met with marine engineer John E. Gowen and his partner, Thomas F. Wells. Wells and Gowen could hardly have avoided following the news accounts of the efforts to raise the *Erie* and the *Atlantic*, since they were reprinted across the nation. Gowen had used an adaptation of Albert D. Bishop's derrick during his internationally acclaimed raising of the USS *Missouri* at Gibraltar. He was aware that Bishop's huge derrick had failed on Lake Erie due to an act of nature— blown down by an unexpected gale. Green assured Gowen that the hardware had been properly applied, and could easily be fastened again. Green agreed to work for the Boston men that coming year on several Lake Erie projects, the first of which would be the *Erie*.[2]

Wells and Gowen had other salvage projects under consideration for 1854. In April, they agreed to terms with marine insurance underwriters for raising the steamer *Humboldt*, which sank near Halifax, Nova Scotia, and the *Staffordshire*, which went under the waves near Cape Sable Island, Nova Scotia.[3] To manage the work, Wells and Gowen assigned an agent, Isaac Coffin, to handle operations on Lake Erie. Green was instructed to prospect for the exact locations of two locomotive engines that had dropped off a barge near the mouth of Grand River, Ontario. Green walked out onto the ice in late February,

1854, and found the engines about a mile and a quarter offshore, in twenty-two feet of water. He took reckonings to relocate the spot later in the season.[4]

The partners were anxious to start their operations before navigation opened on the lake in order to establish their salvage claim. Green began work in late April, 1854, marking the wreck of the *Erie* and guarding the site.[5] By May 10, he had been joined by a veteran British diver employed by Gowen named John Tope. Together Green and Tope worked at securing chains around the hull.[6] Meanwhile, trusses to lift the *Erie* were under construction about thirty miles away in Buffalo. During the last week of May, Green returned to that city to check on the readiness of the truss work. While Green was in Buffalo, Tope and three dive tenders took a boat out from Silver Creek. They did not go out all the way to the *Erie* wreck site, but stood off from the shore when they reached water about forty feet deep. Tope wanted to test some new fittings on his submarine armor.

Tope descended to about twenty feet, but immediately signaled to be hauled up. He complained that his exhaust valve was not allowing air to escape. He went down a second time, but again jerked on his signal line to have them haul him up. Tope took off his helmet and examined the exhaust valve. He removed a spring that he believed was preventing the valve from opening and fitted an exhaust hose over it. Exhaust hoses had not been used by divers for many years, but Tope thought it would work. Then he put the helmet on and prepared to go down again. His tenders lowered him, but when he reached thirty feet, they felt that the line was too heavy, indicating that his suit had little buoyancy. The men above tried to signal Tope, but received no response. Quickly they hauled him up and loosened the bolts on his helmet. Inside, they saw that Tope was dead, and horribly disfigured.[7]

Once Tope's suit had been pumped with air as he descended, the exhaust valve opened and then did not shut, allowing all the air in the suit to rush out through the exhaust hose. Tope then suffered the terror of diver's squeeze, much as William McDonnell had in the same area a year before—the pressure differential clamped the suit to his lower body, causing blood to be forced to his head. He may have asphyxiated, if the squeeze alone didn't kill him. His assistants found his head purple and swollen, with blood still coming out of the nose, eyes, ears, and mouth. His lower body was alabaster white. The body was brought to Buffalo, where Thomas Wells had it sealed in a metal coffin and shipped back to Boston. It was reported that Tope left a wife and four children.

Despite the fatal accident, Green continued work on the *Erie*. By early July all was ready to make the attempt to raise the wreck. Gowen's new truss work was positioned between two ships over the site, and the hoists slowly pulled the chains upward. Gowen's method did not have the lifting power of a towering derrick, but it was simpler, more stable, cheaper, and proved to be sufficient. As the *Erie* was raised fifteen feet off the bottom, the cylinder, a main section of the massive engine, broke away and sank to the bottom. Green went in to adjust the fastenings, and the lifting resumed. This time the hull broke off due to rotted timbers; three-fourths of the vessel returned to the lake bed. The remaining quarter was towed to the mouth of Cattaraugus Creek near Silver Creek, New York.

It was in that harbor that the *Erie* tried to exert its curse one last time. In about forty feet of water, not far from where Tope had died, Green found he needed to adjust the chains under the section they had in tow. While diving under the hull piece, his lines became entangled in the chains. He could not signal the surface—his signal line was also caught on something. Green cut the signal line, and found the free end and tied it to his wrist. He then realized that his air hose and headline (which was used to haul him up) were also trapped. Green cut the head-line, freed it from the chains, and reattached it to his body. Now only the air hose remained tangled. He signaled to be hauled up, even though he knew the men above would only be able to raise him as far as the air hose could stretch from its knot. When the air hose grew taut, Green slashed it through, and in a moment was hauled up out of the water.[8]

The piece was secured and towed to Point Albino, near Buffalo, and was found to contain almost nothing of value. Once again Green returned to the site where the remaining three-fourths of the *Erie* lay. Strong chains were again fastened, and the derrick lifted the 500-ton section off the lake bottom. In a triumphant moment, John Green towed the *Erie* to Buffalo on August 9, 1854—thirteen years to the day after she had sunk. The effort to raise the *Erie* had cost Wells and Gowen about $12,000. When the *Erie* was torn apart at a Buffalo shipyard, the amount that was announced as recovered as of September 1, 1854, was between $15,000 and $18,000, far short of the $100,000 the project had hoped to retrieve.[9]

Green did not wait in Buffalo while the *Erie* hulk was mined for machinery and precious metals. As soon as he had delivered the wreck to the shipyards on the Buffalo River, he informed Wells and Gowen that he was opting out of contract work for any further salvage that season. In this case, it was not fear from his recent close call that halted his efforts, nor was it a business dispute with Wells and Gowen. Green made a quick journey to Boston, where on August 24, 1854, he married Grace A. Jennings of Chelsea, Massachusetts. Green later wrote of her, "In all my diving for treasure I never found one as valuable as this." She was the daughter of Philip Jennings, Sr., a Chelsea shipwright, and sister to Philip Jennings, Jr., also a shipwright and foundryman specializing in marine fittings. Green had probably first met Grace Jennings while consulting with the father or brother earlier that year, when Green had last visited Boston. A column on their marriage record indicated that it was Green's second marriage—a confirmation of the fact that it had been his first wife who was lost on the *City of Oswego*.[10]

Isaac Coffin, the Buffalo agent of Wells and Gowen, continued salvage work through the fall of 1854. In November, he used divers other than Green to raise the machinery of the steamer *Alabama*, while lay near the lighthouse in Buffalo. As late as December, 1854, Coffin informed Buffalo newspapers that it was the intention of Wells and Gowen to construct a specially designed salvage boat, similar to those used by James Eads on the Mississippi, to be used to raise the *Atlantic* in the 1855 season.[11] However, it appears that John E. Gowen decided that Lake Erie salvage was not sufficiently profitable. He and Thomas F. Wells mutually agreed to end their partnership on January 1, 1855.[12] Their agent, Isaac Coffin, continued in the Lake Erie wrecking business out of Fairport, Ohio, for several years, but there is no record that either Gowen or Coffin continued with any ambitions to tackle the *Atlantic* wreck.

John Green and his new bride remained in Boston during the winter of 1854–1855. During that time, Green ordered two new submarine armor suits with his own modifications. The custom features Green specified were a departure from those used by other Gowen divers, James Whipple, and the Siebe English design. Green had the escape valve placed higher up on the helmet in the belief that the warmer, breathed air rose up into the helmet (although carbon dioxide is heavier than unbreathed air). In addition, he thought that the breastplate was too small, and allowed the rubber suit to constrict the lungs, so he specified that the breastplate be extended down as far as the waist. The effect was a heavier armor, which is what he wanted. He moved the intake valve from the helmet down to the extended breastplate to keep the suit better inflated. This made some sense, and in the remaining decades of the nineteenth century, helmets with intake valves on the breastplate were a common variation.[13]

While the new armor was being constructed, Green received an offer from a group of Boston entrepreneurs known as the Boston Sub-Marine and Wrecking Company. This was the same group that had contracted with James Whipple to hunt for the treasure of the *San Pedro de Alcantera* near Venezuela. They invited Green join an expedition seeking treasure from shipwrecks lost in Silver Bank, an area off the coast of the Dominican Republic. In particular, they sought the wreck of a British warship that was rumored to have sunk a century earlier while carrying 150,000 pounds sterling in payroll for troops in the West Indies.

The Boston group of investors had several projects lined up for 1855: the Silver Bank wreck; the wreck of the barque *Wallace* off Grand Manan, New Brunswick; and continued work on the *San Pedro de Alcantara*.[14] Since they were assigned to different projects, there is no evidence that their two most acclaimed divers, John Green and James Whipple, ever met one another. The shifting relationships between teammates and rivals was to get even more complicated within two years, when the renamed Boston Relief and Submarine Company would be pitted as bitter competitors against John E. Gowen.

The Silver Bank expedition left Boston in early spring, 1855. In John Green's written account, he names the wreck they sought as the *Sovereign*, and states that it sank in 1773. However, records place no British ship of that name in that area at that time. A man-of-war, the *Stirling Castle*, was lost on Silver Bank during the Great Hurricane of 1780, but there is no indication that it carried any vast treasure. Green also mentioned that the wreck of a merchant ship, the *Alabama*, was in the same vicinity, and he said it had been lost in 1832. Again, there is no record of this vessel. However, Green and his companions did find the remains of several wrecks on the dangerous reefs. Green was startled by the lack of timbers, which had been devoured by worms. The remaining metal objects they found were encrusted in coral; and most iron artifacts crumbled when touched.[15]

As a treasure-hunting venture, the expedition was a failure. They found little precious metal; extracting many of the objects they found would have required blasting coral; and their electric detonator did not work properly. Moreover, the concession to search for the British wreck that they thought they had received from the U.S. consul to the Turks and Caicos appeared to be overturned by higher authorities. The Boston vessels were ordered by British officials to desist their salvage activities. To avoid returning home empty-handed, they resorted to sponge collecting.[16]

However, Green did bring back something of great value: the experience of diving down to a coral reef and seeing the beautiful flora and fauna of that environment. Green may not have been the first apparatus diver in the Caribbean, but he was the first to describe its wonders. A few years after the journey, his impressions were reprinted in newspapers around the country:

> Here is one of the most beautiful sights imaginable. The water varying in depth from ten to one hundred feet, and so clear, that a person can see under water 300 yards with ease. The bottom, in some places, is as smooth and white as a marble floor. In others it is studded with white columns, from fifty to sixty feet in height, and from six to eight in diameter, resembling the ruins of some ancient palace. Then, again, the corals will be seen growing up like forest trees, branching out at the top, forming arches. There can be nothing more grand than this, when seen from the bottom. To look up through this sublime work of nature imparts a greater joy than the sight of anything on the land.[17]

Green also was overwhelmed by the animal life on display in the coral reef:

> The fish which inhabited those Silver Banks, I found as different in kind, as the scenery was varied. They were of all forms, colors, and sizes—from the symmetrical goby, to the globe-like sun fish; from those of the dullest hue, to the changeable dolphin; from the spots of the leopard, to the hues of the sun-beam; from the harmless minnow, to the voracious shark. Some had heads like squirrels, others like cats, and dogs; one of small size resembled a bull terrier. Some with short blunt noses, others with bills protruding feet beyond their heads. Some darted through the water like meteors, while others could scarcely be seen to move. . . .

The sun-fish, saw-fish, star-fish, dolphin, white shark, ground shark, blue or shovel-nosed sharks, were often seen. There were also fish which resembled plants and remained as fixed in their position as a shrub. The only power they possessed was to open and shut when in danger. Some of them resembled the rose in full bloom, and of all hues. There were ribbon-fish, from four to five inches to three feet in length. Their eyes are very large, and protrude like those of a frog. Another fish was spotted like the leopard, from three to ten feet long. They build houses like the beaver, in which they spawn, and the male or female watches the ova until it hatches. I saw many specimens of the green turtle some five feet long, and, I should think, would weigh some four or five hundred pounds.[18]

The year that John B. Green spent from August of 1854 to August of 1855 surely must have been his happiest: he had recovered the steamship *Erie* where countless other efforts had failed; his fame as a diver reached across America; he had married again; he had helped design new diving apparatus; and he had seen nature at its most glorious, in the seas of the Caribbean coral reef. But Green also knew himself too well. In concluding his adventure to the tropics, he noted with self-awareness:

It is said that there is nothing on earth to which man will so eagerly cling, as to gold. That it occupies his thoughts by day—his dreams by night; and true to this passion, that wealthy safe at the bottom of Lake Erie had been rife in my mind, during all my adventures on the Silver Banks.[19]

Chapter Eleven

~

Race to the *Atlantic* (August–December 1855)

Just as John Green was returning to Buffalo from his journey to the Silver Bank, about thirty miles east of the city a supernatural mystery was unfolding. On a mid-July evening, five men and two boys ventured out onto Silver Lake in Wyoming County, New York, for a night of angling for catfish. The night sky had a few passing clouds, but otherwise the starlight was strong enough that both shores were visible to those in the boat. At about 9:00 p.m., the man sitting in the stern of the boat alerted the others to a long shape floating in the water about a dozen yards away. It appeared to be a long log, between eighty and a hundred feet long and trimmed of any branches, floating on the water. After about a half-hour, the object disappeared, but it reappeared a few minutes later about twenty yards away. The same man who had first observed it stared at it again, and then exclaimed, "Boys, that thing is moving!" They all looked. "See, it is bowing round!"

As they stared, the log shape moved toward them. At fifteen yards they could see its head leaving a wake as it swam at them. In an instant the men hauled up the anchor and put the oars in, hoping to get to shore as quickly as possible. The creature ducked under again, and resurfaced on the other side of the boat, just five yards away:

For the fourth time, when the party were within 35 or 40 yards from their proposed and now nearest landing point, the south side of the inlet, the serpent, for now there was no mistaking its character, darted from the water about four feet from the stern of the boat, close by the rudder paddle, the head and forward part of the monster rising above the water eight or twelve feet in an oblique direction from the boat! All in the boat had a fair view of the creature, and concur in representing it as a most horrid, repulsive looking monster. All agree as to the length exposed to view. . . . When the forward part descended upon the water, it created waves that nearly capsized the boat. . . . The party reached the shore in safety; but frightened most out of their senses.[1]

As word spread through the nearby village of Perry, New York, the local residents decided that some action was needed to rid Silver Lake of the monster. A bounty was put on its skin at a dollar a foot. An experienced whaler was brought it to cruise the lake with harpoon in hand. Another boat full of riflemen traversed up and down the lake. A cabal of local entrepreneurs raised capital stock of $1,200 to capture the monster and exhibit it. This group sent a committee to Buffalo to consult with the expert mariners there for a recommendation as to the best method to trap the serpent. The fishing captains there offered the men from Perry one suggestion: John B. Green, the nation's most accomplished underwater man, was the man for the task.[2]

Nothing resulted from the Perry company's approach to Green. The only reason to recount the tale is to point out that the seasoned tars of Buffalo were having some sport by pointing the Perry residents to Green, their joke being that Green was no more likely to catch the "Silver Lake Monster" than he was to bring up the safe from the *Atlantic*. In their estimation, both were fool's errands.

Green's venture to raise the safe was his own enterprise; he had no partners and was not working under a contract for someone else. In early August, 1855, he chartered the schooner *Yorktown* and a crew of fourteen men and headed out of Buffalo for the short journey to Long Point, the site of the wreck. No markers remained from previous salvage attempts, so they spent several days taking soundings before locating the wreck. They then used lines with grappling hooks to mark the bow and stern of the wreck with buoys. Green began a series of dives to locate the deck and cabin where the safe was located. His new armor performed well, but by the time he reached the upper decks of the ship at 145 feet below the surface, he was working in darkness. The technology of an underwater lamp had not yet been made practical— an understandable challenge, given that all illumination on land still relied on candles and oil lanterns. Green had to grope his way along the *Atlantic*'s riggings, always wary of his lines getting tangled or torn on the wreckage.[3]

The ship itself had shifted a quarter-turn since he had first descended to the wreck three years earlier. Through trial and error, over five or six days of consecutive dives, he was able to consistently land on the wheelhouse, and from there get down to the deck where the cabin with the safe lay, three windows back. On August 23 he reached his arm in through that third window and placed his hand on the safe. He later recounted: "Then I cried out [in his helmet], my God, I've got it! I'm a rich man! And I wept down there in the waters, I was so glad."[4] He signaled to be hauled up to get a line to mark the safe's location. He was lowered again and made one line fast to the railing opposite the cabin window. One buoy on the line reached the surface; a second buoy was set ten feet lower in case a storm or a passing vessel carried away the top buoy.

Green found that each dive to the *Atlantic* produced strange sensations. He later described the discomfort: "While down, I experienced much pain from the pressure of the water. A sensation was constantly felt between the eyes, as if a needle was piercing there; sparks of various colors flashed before my eyes, and the drowsiness, as in my former attempt, constantly attended me, all of which was occasioned by the insufficiency of air in the armor, as, with our pumps, it was almost impossible to force the air down to such a depth."[5] Following the two dives in the morning, Green rested and ate a good lunch with his helmet off, but still in the suit, though it was a hot day. Then he dove a third time, taking with him a long iron prod and a saw. He cut away the cabin wall down from the window and was able to drag the safe out onto the deck walkway, between the cabin and the railing. From that position, the safe could be lifted. Green signaled to be hauled up so that he could get a rope and hook. His dive to move the safe had lasted forty minutes, at about 150 feet.

The lifting tackle for the safe was not yet ready when he was lifted onto the boat. Since he had to wait, Green had his tenders remove his helmet's faceplate to get some fresh air. That last sudden air pressure differential when his faceplate was opened was the moment that John Green felt something wrong with his body. A sharp stabbing pain hit his legs, and in another minute coursed through his whole frame. He fell from his sitting position and lay prostate, unable to move his limbs or muscles. His crew stripped him and, with no idea what was causing his distress, rubbed his limbs with brandy and pepper. Captain Patterson brought the schooner to the nearest town, Port Dover, Ontario. There Green lay for two weeks, with two attending doctors giving him little chance to survive.[6]

John Green, and everyone else at that time, was unaware of decompression sickness, popularly known as the "bends." Gasses breathed under high pressure are dissolved in body tissue, much like gasses in sealed carbonated drinks. The amount of gas saturated in body tissue increases the deeper divers go and the longer they stay at depth, and residually remains after successive dives. If a diver ascends too quickly, the amount of gas coming out of tissue is too great for the body to diffuse through the lungs. The released gasses, like bubbles in an unsealed carbonated drink, are formed within the body. In the mildest cases, this produces a skin rash, but in more serious cases the tissue around joints is distended. In the worst cases, the spinal cord is damaged and paralysis results. Susceptibility to decompression illness varies slightly from person to person, and for reasons not yet discovered, even the same person may be more prone to it from day to day. No human body is immune.

Green may have not been the first diver to suffer from decompression sickness, but he was the first to be so profoundly injured by it. By 1855, it was known that miners working in compressed air for long periods suffered from joint damage, but what was first known as "Caisson Disease" would not be formally identified until years later, during construction of the Eads Bridge in St. Louis and the Brooklyn Bridge in New York. Earlier divers than Green probably avoided decompression sickness because their equipment did not allow them to descend deep enough to cause the condition.

After two weeks in Port Dover, Green's control of his upper body returned. His legs were still paralyzed, but he was able to wiggle his toes. He was brought onto the steamer *Ploughboy* and brought back to Buffalo. In the meantime, newspapers around the country and the globe had reported on his accident, including all the details about how he had found the safe, marked it with buoys, and exulted in the belief that he had attained riches, until coming up the last time. The rapid progress he had made in Port Dover from total paralysis made Green hopeful that in time he would fully recover. However, after ten days in Buffalo, he made no further progress, and was taken back to his home with Grace Jennings in Boston. While lying in Buffalo, Green was quoted as saying, "If God spares my life, I will have that money yet!"[7]

Green's recuperation had two sides. In November, the *Boston Journal* reported that Green had called on them, and that he was recovering rapidly and intended to resume work on the *Atlantic* in the spring.[8] But when Green recalled this period in his autobiography, he said that he laid at home in Boston for five months, unable to take a step. He wrote that it was not until the spring months of 1856 that he was able to stand and walk a short distance with the aid of crutches.[9] There is no question that he intended to return to get the safe; only now, he would need to rely on others to do the diving for him.

≈

By the winter of 1855–1856, none of the nation's leading marine engineers judged that raising the *Atlantic* or diving to recover the safe was likely to prove remunerative compared with other projects. Benjamin Maillefert had begun work in late 1854 on the rock ledges that created falls in the Red River near Alexandria, Louisiana. During low water, those cascades made the river unnavigable. If cleared, shipping would be able to consistently reach Natchitoches.[10]

In January, 1855, Maillefert submitted a bid for new work at Hell Gate, on the rocks known as Diamond Reef. The bidding process was chaotic and politicized, so that even though Maillefert had the proven track record for work at Hell Gate—and a patent on the blasting technique—the contract went to a relatively new firm, the partners Peter V. Husted and Julius B. Kroehl. Maillefert was disappointed, but not idled.[11] His work on the Red River continued into the summer of 1855, whereupon Maillefert immediately began work on a contract for the Canadian government to improve a channel in the St. Lawrence River. The object of that effort was to remove rocks creating rapids between Prescott, Ontario, and Montreal. The quote Maillefert had offered for that work was $750,000, so it is little wonder that the *Atlantic* recovery paled in comparison.

≈

The *Marine Cigar* turned out to be the last submarine that Lodner Phillips built. The precise date of its demise is unknown, but Phillips later claimed it was "in constant use up to the year 1855."[12] The last public mention of it had been in a Buffalo newspaper in late October, 1853, and James Whipple's crewman Daniel Driscoll reported in a letter that a diving boat was in that city in March, 1854. Phillips took out a loan from a Michigan City businessman in April, 1854, for $54.88, payable in sixty days. Since Phillips had borrowed money during the construction of the *Marine Cigar*, this further loan might be an indicator that the vessel was still being worked on at that date. Once the *Marine Cigar* was lost, Phillips lost any designs he had had on taking it down to the *Atlantic*.

≈

Albert D. Bishop, following the loss of the giant derrick over the *Erie* wreck in September of 1853, returned to his home in Brooklyn and continued in the wrecking business. In early 1855 he was hired to raise the recently capsized brig *Rush* in the Hudson River, an operation that succeeded. In June of 1855, Bishop undertook to raise a historic artifact: a section of the "Great Chain" laid across the Hudson River to block British warships at West Point during the American Revolution. Though popular lore maintained that long sections of the iron links rested on the bed of the Hudson, in reality the chain had never been abandoned in the river. Much of it had been retrieved and melted down, and small sections and individual links were already on display at West Point and in other collections. Bishop did not make his effort in vain, however; he did raise a section of the log boom that was used in conjunction with the Great Chain. That boom section is currently on display at Washington's Headquarters State Historic Park in Newburgh, New York.[13]

≈

Bishop's fellow New Yorker, Henry B. Sears, had spent the years between 1852 and 1854 perfecting his Nautilus diving bell, which was a hybrid between a ship-suspended diving bell and an independently moving submarine. The Nautilus had not been ready to deploy during Sears's initial visit to the *Atlantic* in 1852, but Sears made steady progress. In November, 1853, he demonstrated the Nautilus to several high-ranking naval officers at the Brooklyn Navy Yard.[14] A year later, Sears invested heavily in a pearl-fishing expedition to the Pacific coast of Central America. Three of Sears Nautilus diving bells were loaded aboard the barque *Emily Banning* in Brooklyn.[15] Sears did not accompany the company, but he was not idle. In January, 1855, the following notice appeared in several New York newspapers: "To Submarine Operators: The Nautilus Sub-Marine Company hereby give notice that they are now ready to grant rights for the use of their valuable machine, for every kind of sub-marine work, viz: The Recovery of Sunken Property, The Laying of Piers and Breakwaters, Telegraph wires, water pipe, etc. The very great advantage of their locomotive diving bell over other means for operations required to be carried on under water, will read-ily be shown on application at the office of the company, 31 Broadway, H. B. Sears."[16]

The *Emily Banning*, with three Nautilus bells aboard, did not sail directly for the Pacific. Instead, it moored off the coast of Venezuela and commenced operations on the *San Pedro de Alcantera* wreck site. The *San Pedro* wreck had been worked by diver James Whipple and the Boston Sub-Marine Wrecking Company in previous years, but the captain of the *Emily Banning* was reported as having obtained a separate grant from the Venezuelan government. The ship stayed over the *San Pedro* for two months, and sent back letters to America reporting that they were "shoveling dollars" and that the diving bells were perform-ing admirably.[17]

In December of 1855, Sears read a paper in front of the American Geographical Society extolling the advantages of the Nautilus diving bell. Unknown to Sears, the expedition that he had so heavily invested in had reached the Pacific coast of Mexico. There, the *Emily Ban-ning* experienced some sort of distress and put into port in Acapulco, Mexico. The Mexican authorities viewed Americans as interlopers, and threw the captain and crew in prison as "filibusters," that is, mercenaries in service against the government. There the men of the *Emily Banning* spent several months, in deplorable conditions, while Washington diplomats negotiated their release. They were freed with their ship in April, 1856. The diving bells were still intact, but any riches they had collected from the *San Pedro* were gone, and they never had the opportunity to go after pearls. In May, an ad in a San Francisco paper, the *Daily Alta California*, put the barque *Emily Banning* up for sale, along with its three Nautilus diving bells.[18] Sears did not have the money to fund the ship's journey home.

~

Like Henry Sears, the entrepreneurial John E. Gowen had turned his attention from Lake Erie wrecks to the Caribbean. Gowen, probably disappointed by the meagre returns from salvage efforts on Lake Erie in 1854, resumed the import and shipping business he conducted with Thomas F. Wells. However, they ended their partnership on January 1, 1855, in a public notice appearing in Boston newspapers. Wells continued to operate the submarine armor manufacturing. Gowen, on the other hand, was intrigued by the huge profits to be made from the miracle fertilizer, guano. Guano, the dried excrement of seabirds, was found in large quantities on islands in the Caribbean and off the Pacific coast of South America. It had recently been found to produce dramatic increases in farm output due to its rich nutrients: nitrogen, potassium, and phosphate.

The 1850s saw a "guano gold rush," with American businessmen scurrying to locate isolated deposits off of Central and South America. The frenzy culminated in the Guano Island Act of 1856, passed by the U.S. Congress to assert that any deposits found by U.S. citizens on unclaimed islands could be possessed by those prospectors. The law was passed to benefit Gowen and other assertive American speculators like him. Gowen, through his shipping contacts, had learned of huge guano deposits on the Los Monjes Archipelago, a group of rock specks claimed by Venezuela, though Gowen was not immediately aware of that claim. After spending the first eight months of 1855 raking in tidy profits from these guano mines, a rival American company pointed out to the Venezuelans that Gowen was encroaching on their sovereign territory. Under threat of expulsion, Gowen agreed to hand over the operation to the rival company after fifteen months, and during that time would pay them a share of the profits.

Stifled at Los Monjes, Gowen used the Guano Island Act to lay claim to another rich deposit on Sombrero Island in the Lesser Antilles. This time, rather than setting up his own mining operation, Gowen sold half his interest to a New York company and let them operate it. In spite of the pecuniary charms of mining bird excrement, after 1855 Gowen was once more attracted to the special challenge of underwater engineering; his next venture in that pursuit would take place far from Lake Erie, in a different hemisphere of the globe.[19]

≈

The protégés of George W. Taylor's diving franchise, Charles B. Pratt and James A. Whipple, spent 1855 engaged in separate long-term projects. Pratt was still, after four years, at work in the treacherous currents of Hell Gate trying to recover the fabled treasure of the HMS *Hussar*. Numerous critics of Pratt's efforts pointed out that British authorities had stated that the payroll money had been offloaded before the *Hussar* sank, but Pratt knew that the British themselves had mounted a couple of expeditions to salvage the *Hussar*, so he reasoned that they must have had a good reason to do so. Tides only allowed Pratt to dive two hours a day, at most, which is one reason why the operation was so prolonged. In July, 1856, during Pratt's sixth season working over the site, a Boston paper noted that all that had been found so far were a few cannons and ship furniture.[20] Pratt appeared to view this as a hobby rather than a vocation, and spent the majority of his time conducting nondiving business in Worcester, Massachusetts.

James Whipple had contracted with the U.S. government in December of 1854 to supply submarine armor for divers and to assist with clearing the harbor and constructing a sea wall and dock at Pensacola, Florida. The main obstacle that Whipple faced was the removal of a dry-dock caisson that had been constructed at huge expense thirty years earlier, but that was obsolete by the 1850s due to its size. Whipple used two tons of gunpowder to remove the caisson. He then helped construct a granite sea wall, shielding slips where new steamer warships were to be built.[21] However, none of Whipple's impressive work at Pensacola excited much public interest. For a period of several years, the one diving venture that dominated national attention was the effort to recover the safe of the *Atlantic*.

~

Martin Quigley's diving career took a track even more unheralded than Whipple's. In early 1854, following his break with former partner John Green, Quigley moved to Detroit and secured a contract with the owners of the steamer *Minnesota*, which had sunk in shallow waters near Amherstburg, Ontario, where the Detroit River empties into Lake Erie. The *Minnesota* had hit a rock, opening a gash in hull her that was thirty-five feet long and three to nine feet wide. Moreover, when the ship sank, the side with the hole settled down onto the muddy bottom. Quigley labored day after day to excavate under the hull to access the hole and patch it. It was strenuous, dirty work, done largely within the submarine armor but lying on his back on the river bed, working with no light.[22]

His herculean efforts were praised in the Detroit papers, but drew little attention elsewhere. By the next season, 1855, Quigley had an even larger contract to raise the propeller ship *Princeton*, which sank in April, 1854, in Lake Erie near Quigley's former home, Westfield,

New York. The *Princeton* had carried 200 tons of freight merchandise valued at $200,000 and went down in sixty feet of water after colliding with ice floes.[23] For this salvage job, Quigley hired extra divers, including Elliot P. Harrington. Like much commercial diving, this work was unglamorous and unheralded. It was risky work, but to the public the danger was unseen and banal.

Harrington had recently moved from Chautauqua County, New York, to the Detroit area. His wife, the former Emaline Maxon, was raised in Ypsilanti, just west of Detroit. Harrington's sister-in-law, Caroline F. Maxon, married a man named James H. Phillips. James Phillips became Elliot Harrington's partner and main diving tender.[24] Later in May of 1855, after working with Quigley on the *Princeton*, Harrington was hired to recover bodies and fixtures of the propeller ship *Oregon*, wrecked near Hog Island in the Detroit River.[25] In June of 1855, Harrington and Phillips used their submarine armor to search for the body of a drowning victim in a small lake near Ypsilanti.[26]

After the shipping season closed on Lake Erie in the fall of 1855, Harrington took a job working with wrecking maven James Eads in St. Louis. Harrington served on the diving bell boat *Submarine No. 4*, the newest and most sophisticated of Eads's wrecking barges. *Submarine No. 4* was equipped with diving bells and Gwynne centrifugal pumps, capable of extracting enormous amounts of sand and silt from around a sunken vessel. In early December, 1855, the steamer *Young America* and a barge it was towing sank near Alton, Illinois. *Submarine No. 4*, with Harrington working as a diver, recovered $4,000 in freight from the wreck.[27]

It is not known how much of the time Harrington spent with Eads was involved in working from diving bells or in submarine armor. Many years later he told an anecdote about taking two women down forty-two feet deep in a diving bell and cooking an oyster dinner for them while underwater (Harrington was married at the time).[28] How long Harrington stayed in St. Louis is unclear, but he was probably there in late March, when *Submarine No. 4* and a dozen other steamers were destroyed by the Great Ice Jam of 1856. Of that event, the *St. Louis News* wrote:

> We have never witnessed a spectacle of more appalling grandeur than this breaking up of the ice. It was brought about by a rise, of some twelve or fifteen feet, and not by rotting of the ice. The ice was hard as ever, ranging from ten to twenty inches in thickness; and as it moved slowly down, sweeping everything like chaff before it—snapping stout chains and cables like whip-cord, and roaring like thunder—the scene was grand and terrific, and impressed upon the minds of the ten thousand spectators the resistless might of the elements of nature, and the impotency of human fabrications and contrivances.[29]

Though he lost *Submarine No. 4*, Eads had other diving bell steamers, and the many wrecks created by the ice jam meant a boom in business for his wrecking company. However, Harrington did not remain in St. Louis to take advantage of the plentiful wrecking jobs. When shipping opened on the Great Lakes in April of 1856, there was a collision on Lake Huron involving the steamers *Northerner* and the *Forest Queen*. The *Northerner* sank in an area so shallow that the upper deck stuck out of the water, but within a week it was washed away by the waves. Harrington, back from the Mississippi, was hired to recover the valuables from the lower decks.[30]

None of the diving that Quigley, Harrington, or any others employed by Quigley did in 1854 and 1855 took place in waters over sixty feet in depth, nor did any of the wrecks they operated on contain treasure other than valuable freight. A century later, they might have been called "blue-collar" divers. By 1855, their workmanlike approach to diving was becoming commonplace, despite the lack of headlines or the promise of instant riches. In job after job, they built a résumé of successful but unheralded exploits.

Chapter Twelve

≈

The Safe Recovered (1856)

John B. Green meant to make good his vow to recover the safe from the *Atlantic*, in spite of his disability. By late spring, 1856, he was able to walk with the assistance of crutches or a cane on solid ground. More doubtful is whether he could move at all on the deck of a ship. He convinced two men with some diving experience to venture out to the site from Buffalo in late June, 1856. On board their chartered ship was Green's brother-in-law, Theophilus Peter Tremble. Tremble had married Green's sister Mary the year before, in 1855. Tremble and Green may have been acquainted long before that time; a Tremble family anecdote relates that Tremble, who was nine years older than Mary Green, saw her as an infant when their families were boarding together and told everyone that he would marry her someday.[1]

Later in the 1860s, Tremble's census occupation was listed as "submarine diver," but it is not known whether Green asked Tremble to go down to the *Atlantic*, or whether he was merely serving as a tender when they arrived at the wreck site on July 1, 1856. It would be interesting to know what John Green said to the divers he brought. He expected them to finish the job that he had started, though he surely believed himself to be the most capable diver alive. Did he prepare them against the same discomfort he had experienced on his descent? Did he offer a financial incentive if they succeeded? Had he modified the diving apparatus to perform differently?

When Green and his crew arrived at the *Atlantic* site off Long Point on July 1, they could not find the marker buoy left behind ten months earlier. This was not unexpected; buoys often were swept away during the winter by ice or were caught by passing ships during the shipping season. Still, it annoyed Green that he would have to waste time dragging a line to pinpoint the wreck again, and then his divers would need to orient themselves to locate the deck aisle where he had pulled the safe. Once the site was rediscovered, Green sent a diver down. However, the man signaled to be hoisted up after descending just sixty feet, less than halfway to the wreck. Green sent his second diver down, but he made it no deeper than the first.

On July 4, Green was thoroughly disgusted with his hired help and put on the submarine armor himself, in spite of his disability. Green made it all the way down to the wreck, though he felt weak and knew he could not stay down very long. He was able to orient himself quickly, and reached the deck aisle where he had left the safe. In darkness, he felt around, his hands only touching the rail on one side and the cabin wall on the other. The safe was gone. He later wrote, "I felt for the safe again. It was gone. All—all my efforts for nothing! Never—never did I rise to the surface with so heavy a heart!"

After Green was brought up onto the deck of the ship, the face plate was opened on his helmet. Green was unable to speak, but motioned for his tenders to take his armor off. As soon as that was done, he fell to the deck paralyzed once again, and lost consciousness. His crew brought him back to Buffalo, where he was treated in the same manner as he had been a year earlier. It was there that Green learned who had recovered the safe.[2]

≈

Martin Quigley, Elliott P. Harrington, Charles O. Gardner, and a Detroit diver named William Newton had arrived at the *Atlantic* wreck on June 18 in the schooner *Fletcher*, two weeks before Green and his crew arrived. Whether or not they were aware that Green was preparing a return expedition to the *Atlantic* is not known, but Green had publicly stated that intent during the past months. Moreover, although the nation's leading marine engineers and divers had shied away from further work on the *Atlantic*, other local wreckers still had their eyes on the prize.

One aspirant was Miles Osborn of Conneaut, Ohio. Osborn had a patented method for raising ships that employed arrays of large, air-filled white-ash barrels, similar to Taylor's or Irwin's rubber camels.[3] Osborn invented a crank mechanism that would thrust the barrels down a chain, increasing their lifting power. Osborn's plans were published in newspapers in June of 1856, but were probably already no secret to Lake Erie mariners.[4] The idea of sinking wood barrels to great depth would have been doomed to failure, but Quigley, Harrington, and Co., would still have perceived Osborn as a competitor.

Harrington, using Wells and Gowen submarine armor, made all the dives down to the *Atlantic* over the next week, staying only a few minutes during each dive. The conditions Harrington later described were very similar to those reported by Green: visibility was almost nonexistent below fifty to seventy feet; the deck of the Atlantic was covered in three to four inches of sediment; his hands grew very numb; he could feel blood rushing to his head, and saw bright flashes in his helmet, like electric sparks.

However, in the critical details, the accounts of Green and Harrington diverge. Harrington never mentioned seeing any marker lines or buoys left by Green. The fact that Green had left buoy lines and had found the safe was no secret—those facts had been published in newspaper accounts of Green's prostration in 1855. Green later stated that he cut around the cabin window and pulled the safe out on the deck aisle. Harrington said that he was the one who broke the window; he then spent an entire day trying to cut an opening around the window, and finally resorted to having the men in the ship above pull the window frame out by attaching it to a line to the ship above. After that was done, Harrington said that he fastened another line to the safe and pulled it out of the opening that had been made in the cabin wall. Once it was free, Harrington's partners on the schooner lifted it to the surface.

The safe was recovered on June 27, 1856, the same day that John Green set out with his crew; the schooners of the two expeditions may have passed each other. The safe was fairly small, measuring twenty-eight inches by eighteen inches by sixteen inches. Quigley, Harrington, Gardner, and Newton opened it and discovered $5,000 in gold, $31,000 in paper bills, and six watches, two of which were gold. Nearly $3,000 of the bills were ruined; a story came out years later suggesting that Harrington's mother tried to dry out some of the bills using an iron, only to have them disintegrate.[5]

When reporting on Harrington's daring success, the *Detroit Advertiser* editors decided that the plain facts were not sufficiently melodramatic. Either that, or someone close to the four men decided to play a prank on the newspaper. The *Advertiser's* embellished account included ridiculous details:

> The upper deck of the steamer lies 160 feet under the water, and far below where there is any current or motion. Everything is therefore exactly as it first went down. When the diver alighted upon the deck, he was saluted by a beautiful lady, whose clothing was well arranged, and her hair elegantly dressed. As he approached her, the motion of the water caused an oscillation of her head, as if gracefully bowing to him. She was standing erectly, with one hand grasping the rigging. Around lay the bodies of several others, as if sleeping. Children hold their friends by the hand, and mothers with their babies in their arms were there. In the cabin the furniture was still untouched by decay, and, to all appearances, had just been arranged by some careful and tasteful hand.[6]

Harrington, it was said, was mortified when this fantastical account appeared in print, not only in the *Detroit Advertiser* but reprinted around the country. To set the record straight, the normally reticent Harrington granted an interview to the *Cleveland Herald*. In it, Harrington offered details on the dates and durations of all his dives on the *Atlantic*. He reiterated that there was no visibility at the depth of the wreck, and therefore no fanciful visions.

As the four men prepared to divide the spoils—close to $6,000 per man—they doubtless felt their fortunes were made. In that period, the average annual salary of a blacksmith or carpenter was $600 a year, so the booty was immense. However, the safe and its contents were subject to salvage law, and so when the four men did not present their spoils to a United States Court, they received a visit from an attorney for the American Express Company. The lawyer offered them a deal in which they would keep $7,000. It was a reasonable compromise, on par with what they might have received through a court considering existing admiralty law. It is possible that with the aid of a talented lawyer they could have eked out a larger share, but they opted to settle immediately. Instead of $6,000 per partner, they received $1,750.

Green, laid up and in great pain, did not handle his misfortune easily. He wrote a letter to the *New York Herald* after that paper reprinted the interview that Harrington had given to the *Cleveland Herald*:

> An article was published in the *Herald* of yesterday, taken from the *Cleveland Herald*, wherein it states that Mr. Harrington (who obtained the safe,) had accomplished a feat never before successful. Allow me to say that this statement is erroneous, and calculated to do me great injustice.
>
> During the summer of 1855 I made thirteen dives to the deck of the *Atlantic*, and in making my last dive was absent from the surface forty minutes; in the meantime I succeeded in finding the location of the safe, and also attaching a buoy to the wreck near the safe.

In making this dive and remaining forty minutes under
water, the result was I became partially paralyzed, and was
obliged to suspend operations for nearly a year; however,
the buoy still remains attached to the wreck, and it was
with the aid of this buoy that Mr. Harrington succeeded
in obtaining the safe.

—John Green, Diver.[7]

For many years, Green repeated this assertion that Harrington had followed the buoy Green had left straight down to the safe. Perhaps that
was true—but it was something that John Green could not have known.[8]
Green also criticized Harrington for many years over the concocted
Detroit Advertiser story of spectral figures on the wreck, even though
Harrington denied that he had ever told such a tale and acted immediately to repudiate it.

Harrington was disinclined to engage in a public spat with Green,
and in fact there is nothing in print to show that Harrington ever mentioned Green by name. Fame was not something Harrington sought;
two weeks after the safe was recovered, his name faded from the newspapers. The whole experience with the recovery of the safe did not
encourage him to go after other treasures, or even to continue much
longer with gritty, everyday salvage diving. He completed a few more
jobs in the 1856 season, and then took his earnings and moved his
family to Charles City, Iowa, where he resumed the carriage-making
business.

Harrington did get one offer in the wake of the *Atlantic*. He
received a letter from Daniel D. Chapin of New Jersey. Chapin, the
tireless prospector who had tried to recover the *Erie*'s treasures a decade
earlier, wanted to recruit Harrington to dive for the strongbox of the
Lexington, the same object that had eluded numerous divers. Chapin
assured Harrington that he could pinpoint the box using his "marine
compass." Moreover, Chapin informed Harrington that he had found

Captain Kidd's treasure ship in the Hudson River, and just needed a capable diver to retrieve its wealth.[9] Harrington ignored the letter. When Harrington left for Iowa, he took with him a gift bestowed by a gracious Henry Wells, president of the American Express Company. It was a gold watch with a cover engraved with a depiction of a diver descending from a schooner, heading toward a wreck representing the *Atlantic*. Inside an inscription read: "Presented to E. P. Harrington by the American Express Company, as a reward for his heroic efforts in recovering their safe from the wreck of the steamer *Atlantic*. On the 19th day of August, 1852, the *Atlantic* was wrecked and sunk in Lake Erie; and on the 22nd [sic] day of June, 1856, Mr. Harrington performed the miraculous feat of diving to the wreck and fastening a cable to the safe at a depth of 170 feet from the Lake surface."[10] The feat of retrieving the safe was rightly called "heroic," and marked the last time that a dive to any wreck in Lake Erie attracted national attention. However, it did not mark the last public actions of either Green or Harrington. As for the word "heroic," the American public's understanding of that concept would be redefined thousands of times in the coming decade.

Part Three

THE AFTERMATH

Chapter Thirteen

~

The Moving Panorama (1857–1860)

Though incapacitated for the last half of 1856, John Green expected to recover from his paralysis to the extent that he had after his first episode a year earlier. With limited mobility, he hoped to once again at least organize salvage operations. With such expectations in mind, he used his dwindling savings to have a schooner built for this purpose. It was constructed by shipbuilder Charles Stevens in Irving, New York, and was named the *Grace A. Green* in honor of his wife. For the summer of 1857, Green outlined a tour of Lake Huron, in the area along the coast from Thunder Bay, Michigan, past Middle Island, and up to Presque Isle. There he hoped to find freight that was frequently thrown overboard by foundering ships.

The voyage to Lake Huron did not prove profitable. None of the divers Green hired were able to live up to his standards, and Green himself was unable to do much diving. He blamed the failed voyage on the men he had recruited. On their way back to Buffalo from Lake Huron in September, 1857, Green had the *Grace A. Green* stop over the wreck of the *City of Oswego,* a site he had already visited time and again during the previous five years. The wreck had been picked over many times by Green and other salvers, but the place still lured him. It is questionable whether anything of physical value remained at the site, but Green may not have stopped there for that reason. More likely, he was looking for something he could not recover.

The *Grace A. Green* lingered too long over the *City of Oswego* site, and was buffeted by a strong wind the night of September 16. Reluctantly, Green headed the schooner to Cleveland. As he reached port and attempted to dock, the gale threw his boat against the unfinished West Pier, where a jutting timber knocked a hole in her hull. Green and his crew jumped off. The *Grace A. Green* was later repaired, but John Green relinquished his ownership.

Writing in 1859, Green avowed that 1857 marked the end of his diving career, but he omitted a few minor incidents. During the winter of 1857, Green formed a partnership with a civil engineer named William A. Kendrick of Boston. Kendrick was mainly devoted to bridge work, wharf construction, dry docks, and dredging; many of these operations employed divers, but Kendrick's interest in wreck salvage appears to have been tenuous. Moreover, newspaper reports indicated that Green would be managing wreck operations in the Great Lakes from Detroit, while Kendrick was to remain in Boston. It is possible that the men met while Green was living in Boston and working for the Boston Sub-Marine and Wrecking Company; or perhaps Kendrick was introduced to Green through his wife Grace and her family.

In April of 1858, Green announced the ambitious schedule of wreck diving he planned for the coming season: the *Northern Indiana*, the *Wisconsin*, the *Chesapeake*, the *Northerner*, the *Charter* . . . and the *City of Oswego* (again!). However, it appears that Green and Kendrick did not have the capital to get the operation underway. In July, Green made a public exhibition of diving in the Detroit River aboard a special excursion of the steamer *Sultana,* with collected funds to be used to start his wrecking business. It does not appear that enough was raised to purchase a vessel, although there is a mention made in the fall of 1858 that "Green's divers" were being taken by a tug to recover steam pumps that sank with the schooner *Malakoff.*

If Green was not very active in the wrecking business through the 1858 season, it may be that he spent that time preparing other means of income. Over a series of three nights, from December 16 through 18,

1858, Green gave an illustrated lecture at the Detroit Theatre describing his diving experiences. As props, Green had a few painted scenes and also brought his submarine armor. The talks were described as benefits, and were scheduled in between the productions of two short plays that were normally on the bill. Most likely, donations for Green were collected from the audience.

Though physically unable to continue diving, John B. Green discovered a different sort of treasure to share—his own story. Over the next few weeks, more painted scenes were added to his talk, and by mid-January, 1859, the illustrations were given top billing—described as the *Submarine Panorama*. By March, 1859, the production was listed as including nine scenes:

1. The sinking of the *City of Oswego* and the burning of the *G. P. Griffith*

2. The burning of the *Erie*

3. The sunken wreck of the *Erie*

4. The *Atlantic* colliding with the *Ogdensburg*

5. The *Atlantic*, partially submerged

6. The sunken wreck of the *Atlantic*, on the lake bed

7. The wreck of the USS *Missouri* in Gibraltar harbor

8. A diving bell, as modified by Green, in operation

9. Splicing together of the transatlantic cable

Panoramas were an established entertainment medium in the 1850s, both as permanent wall installations in specially designed rooms, and as rolls of canvases set up on giant spindles to be scrolled before the audience (moving panoramas). In a sense, they were the motion pictures of that era. In the spring of 1859, other panoramas touring the country included *Banvard's Destruction of Jerusalem, Bean's Panorama of the Overland Route to India*, and *Dr. Beale's Panorama of the Mammoth Cave of Kentucky and Niagara Falls*.

After he left Chicago, advance notices for showings of Green's *Submarine Panorama* stated that it contained *thirty-six* scenes, not nine. The discrepancy could be explained if there were nine canvases, each averaging four scenes—a typical configuration of moving panoramas. If so, the *Submarine Panorama* must have represented a considerable investment on John Green's part, putting it on par with the notable panoramas mentioned above. The cost of the paintings, travel and accommodations, theater or hall rental, and publicity had to have drained his savings—and perhaps he mortgaged everything he had in the expectation of large audiences.

At about the same time Green gave his illustrated talk in Chicago's Mechanics' Institute Hall on March 25, 1859, he had in hand copies of his self-published autobiography, *Diving, or, Submarine Explorations, being the Life and Adventures of J. B. Green*. This forty-two-page booklet was printed for Green by J. S. Leavitt, a Buffalo bookseller. This first edition of his autobiography was probably dictated in the later part of 1858, before Green began his panorama circuit. In the "Conclusion" section of the booklet, Green states, "This brief and somewhat confused collection of facts and reflections, gives but a faint conception of the more elaborate and specific work which I have in contemplation. To portray the enchanting scenes and describe the new and strange objects which the ocean world presents to the enraptured vision of the submarine Explorer, would require a more graphic pen than this small beginning could hope to command."

Green's own voice comes through in the booklet, but it is also apparent that he had assistance with the writing and editing. Green, by his own admission, had limited schooling, yet *Diving* is sprinkled liberally with literary quotes and biblical allusions. Certain passages sound as if they were crafted to impart a moral lesson, and other lines are in the flowery style favored by editors of that era. One example: "The broad expanse of the St. Lawrence, at this place, presents one of the most lovely scenes at sun-set the eye ever beheld. It arises from the gorgeous hues of the clouds, reflected from the clear surface of the slumbering stream."

The strong possibility that Green's two versions of his autobiography were heavily edited by another hand makes the task of deconstructing the text difficult. For instance, after discovering the *Atlantic* safe was gone, Green comments in *Diving* that "Perhaps these sad disappointments were sent as monitors to teach me the fallacy of trusting to those 'riches which take to themselves wings and fly away.'" Many passages like this one express a self-awareness and reflection that contrasts with other sections where he scoffs at the talents of other divers and questions Harrington's integrity. The incident where Green believes his diving partners are cheating him, and decides to cheat them back, is especially jarring, given the philosophical tone found in much of the language of the text.

The sage, reflective narrator of Green's autobiographies must also be measured against troubling reports of his behavior between 1859 and 1860. In Indiana, Green lectured at Crown Point and then moved on to Fort Wayne. There, things fell apart. The editor of the *Fort Wayne Weekly Republican* had a complaint against Green:

> Newspaper editors are notified to be on the look out for one Mr. J. B. Green, who is going about showing Submarine Paintings, and calling himself the World Renowned Diver, etc. He was here for a day or two, and has decamped without paying his bills, or at least without paying some of them.—He left owing us eleven dollars, a carpenters' bill of nine dollars, and whether any others or not we do not know. He left here for Van Wert, Ohio.

Two months later, the Fort Wayne printer was still in high dudgeon:

The *Buffalo Express* of Wednesday morning announces that the friends of Mr. John B. Green, the celebrated Submarine Diver, have tendered him the benefit of a complimentary pleasure excursion, per steamer *Arrow*, to Falconwood [a fashionable resort near Buffalo on Grand Island] to-day.

This same John B. Green was in this city about the last of April, for the purpose of exhibiting a panorama of Submarine Paintings, and on leaving here made an effort to swindle a landlord out of his hotel bill, but was overtaken at Van Wert, Ohio, and his panorama attached and brought back, where it still remains, and he, "the celebrated Submarine Diver," was "tendered the benefit of a complimentary pleasure excursion" by rail—he about as cleverly befuddled with bad whiskey as such impostors generally get. He played a similar game upon others on his route hither, and will doubtless give our Buffalo friends a sample of his *diving* propensities, if they are not careful of their pocket-books.

The trail of the *Submarine Panorama* ends in Fort Wayne, Indiana, where it was held for unpaid bills. A showing in Cleveland had been announced for mid-May, 1859, but was not realized; and it was never mentioned in print again. Today, moving panorama canvases from that period are very rare. One of the cruel ironies of John Green's story is that today, his *Submarine Panorama* would be an artifact of incalculable value, no matter what its artistic merit. Finding it would be worth far more treasure than any sunken specie that Green and his fellow divers sought.

Green returned to Buffalo with his hopes ruined. The latest failure, and perhaps the role that Green's character flaws played in that disappointment, were the last straw for his wife Grace. She had nursed her husband through two incidents of paralysis, and had perhaps been instrumental in his efforts to communicate his story through the panorama and book. The insinuation made by the Fort Wayne editor that Green now had a drinking problem was the first time that condition was attributed to Green—but it would not be the last. Grace A. Green, née Jennings, returned to her family in Boston.

Despite his marital problems, John Green was able to summon his spirits sufficiently to have an expanded edition of his autobiography published in October, 1859. It was retitled *Diving with & without armor: containing the submarine exploits of J.B. Green* and printed by a different Buffalo press than the first edition. It was expanded from forty-eight to sixty-two pages, but that was accomplished mainly by padding the descriptions of incidents already mentioned in the first edition. The editorial hand of the unknown ghost writer is even more apparent in this edition.

Expunged from Green's second edition was any mention of Grace A. Jennings, including the reference in which he said of her, "In all my diving for treasure I never found one as valuable as this." The second edition also is where Green named his treacherous villain "Perglich," in contrast to the first edition, in which all other members of that first diving company went unnamed. As time went by, and with the breakup of his marriage, Green's bitterness seemed to deepen. The new edition made more disparaging remarks about Harrington's account of the recovery of the safe.

Green spent the last months of 1859 and the early months of 1860 traveling around to Lake Erie ports and upstate New York selling his new edition. The steamboat line magnate Eber Ward, owner of the *Atlantic*, was said to have told all his captains to give Green free passage on all Ward ships. Several newspapers ran a notice about his new edition and referred to Green's dependence on sales of it to provide an income. Some newspapers ran lengthy articles with reprints of Green's sections on diving on the Silver Bank, providing American readers with the first descriptions of animal life on a coral reef. To Green's credit, he avoided sensational anecdotes about shark attacks or grappling with octopi, in contrast to other nautical yarns popular at the time. Green mentioned both creatures, but related that they had little interest in divers walking the bottom in submarine armor.

Green's book tour took him finally to his hometown of Oswego, New York. There, his fortunes reached their lowest ebb. The *Oswego Commercial Times* noted his saga as a pathetic parable (without mentioning him by name):

Fickle Fortune—Most of our readers will remember the famous diver and sub-marine explorer of the lakes. The individual was originally a resident of Oswego, and for some years obtained his livelihood by odd jobs and laboring upon the docks. He finally became somewhat noted about town as a diver, and did considerable in the way of diving up coal, old iron, and other articles lost overboard from vessels in the harbor.

From this he engaged in the wrecking business, invented a sub-marine armor, and made himself famous by his operations in Lake Erie, in exploring the wrecks of sunken steamers and recovering the treasures. The diver made a deep dive into the tide which leads to fortune, and it is supposed he secured a handsome competency by his operations.

But it appears the diver could not stand prosperity; he retired from business, and adapted his habits to an aristocratic sphere of life—took to broadcloth, diamond pins, eschewed straight whiskey for smashes and cocktails, and shook a tolerable hand at draw-poker. Such accomplishments are not the inspiration of a vision, and of course too sudden accelerated velocity in one's mode of life is hazardous. It proved so in the present instance, and landed the diver farther on the road to perdition than Morse's telegraph would have taken him in a year.

He makes the entire revolution of the wheel of fortune, and next turns up in the jail in this city, under charge of the poormaster—probably without the requisite capital to accomplish a respectable suicide. The story of the diver is a warning and suggestive to young men about starting in life who contemplate diving.

For John B. Green, who now depended on telling his story to others to survive, this dismissal from his hometown was the cruelest blow.

Chapter Fourteen

≈

War (1861–1865)

The idea that underwater engineering could be used for attacks on ships had a long history in America, promulgated by inventors of submarine vessels and contact mines, known then as "torpedoes." It was recognized that the military role of divers was limited to ship maintenance and removal of obstructions, since divers required a support ship above them at all times. Even so, many of the engineers involved in diving activities also contemplated military uses for other underwater technologies. For example, George W. Taylor had demonstrated torpedo demolitions along with submarine armor all through the 1840s.

When Taylor and Samuel Colt initially tried to interest the U.S. government in the use of torpedoes as offensive weapons, they were rebuffed by many who thought that attack by an unseen opponent was barbaric. While those with a citizen-soldier mentality in the United States military establishment hesitated, criminal minds were more receptive. One of the earliest adopters of torpedoes as weapons was none other than a Lake Erie diver of the 1850s, albeit not one involved in any of the daring dive operations over the great shipwrecks.

In 1859, a Cleveland dock worker and occasional diver named Harrison R. Cooley was arrested for plotting to blow up a Lake Erie passenger steamer. The *Cleveland Plain Dealer* reported:

It is well known to the public that soon after the opening of spring navigation, that enterprising, go-ahead lake captain and boat owner, E. B. Ward, placed the steamers *Sea Bird* and *Arctic* on the Buffalo and Cleveland line of lake travel in competition with the Lake Shore Railroad—and that the Railroad immediately thereafter brought out the steamers *Western Metropolis* and *City of Buffalo*, to run in competition to Ward's Line.

Knowing the feeling of rivalry existing between the two lines, and thinking, doubtless, as many a villain does, that the base cupidity of his own black heart might be shared by others, a wretch named Cooley, who is somewhat known along the docks as an expert diver, approached Capt. Ward two weeks since with one of the most fiendish propositions ever made by one man to another. He proposed to relieve Capt. Ward of the competition of the other boats by blowing them up with a sort of torpedo while they were tied up at the dock in this city, which explosion would sink the boats and render them useless. Capt. Ward's first impulse upon perceiving the drift of the villain's proposal was to denounce and expose him, but upon reflection, he concluded to repress his indignation, appoint a meeting where the conversation could be overheard by a third party, and draw out of him the details of his detestable plot.

Another interview was accordingly appointed. Punctual to the hour the parties met, when the plan, in all its horrible details, was unfolded to the Captain, every word of which was distinctly heard by an officer who was secreted in the room. He had already, he said, taken soundings at the boat's dock, and perfected his plans. For the purpose of avoiding suspicion he would some day when the *City of Buffalo* should be in port take the torpedo into a fishing boat and sail out into the lake apparently for the purpose of fishing; then paddle his way back into the river and while passing the boat drop his torpedo at the right point, fasten the fuse to a spike under the dock, in which condition it would safely remain until the time appointed for the explosion . . . when he would row under the dock, find the fuse, apply fire to it, and then get out of the way, he having ten minutes to make good his retreat. . . .

On Saturday, Marshal Craw, fearing that serious consequences might attend further delay, concluded to arrest Cooley. . . . There is no law for the punishment of men engaged in such a conspiracy, where their purpose is not accomplished. It is to be regretted. Cooley is in the County Jail, and will soon be tried for [the unrelated crime of] counterfeiting, as it is said he can be convicted on that charge.

Cooley was an employee of a Cleveland dockmaster for the Ward Line named William Nelson—the same man who had once hired John B. Green as a diver in 1853.

The episode involving the would-be saboteur Cooley illustrates that competition for passenger traffic on Lake Erie was becoming fierce, not only between steamer lines, but between ships and railroads. Lake Erie was feeling the effects not only of technological change, but of the worldwide economic Panic of 1857. Migration of settlers to the western United States declined sharply, and demand for goods shipped from the West to the East also declined. The downturn signaled the end of the era of the great palace steamers on Lake Erie. When western migration picked up again after the Civil War, railroads had supplanted Lake Erie as the preferred route to the west. Freight traffic that once was carried by the Erie Canal through to the Buffalo area could now be found on iron tracks. The great wrecks of the *Erie*, the *G. P. Griffith*, and the *Atlantic*, which occurred in the space of eleven years and claimed the lives of hundreds of immigrant settlers, would not be repeated in the remaining decades of the nineteenth century.

Many ships continued to founder on Lake Erie and the other Great Lakes, though the tragedies were not on the scale of the three wrecks mentioned above. The massive loss of life that occurred when the palace steamers went down was lessened not only by the decline of big passenger steamers, but also by increased attention to safety measures and inspections on passenger vessels. The craze for setting steamer speed records declined, as did the foolhardy practice of forcing other ships in a lane to make way.

Still, the inevitable collisions, gales, and fires provided steady work for wreckers, and it was to that vocation that Martin Quigley returned after making a modest profit from the safe of the *Atlantic*. From 1857 through 1863, Quigley continued to be the leading salvage diver on Lake Erie and Lake Huron. At some point in the late 1850s, his marriage collapsed, with his wife remaining with her family in Chautauqua County, New York, while Quigley boarded in rooms in Detroit and Buffalo. When Fort Sumter was fired upon in April, 1861, Martin Quigley was fifty years old, too old for the military draft that followed. If he had tried to enlist, he would have been told either that he was too old or that his civilian work was vital.

In the summer of 1863, Quigley was still in Detroit pursuing salvage when a marine news editor of the *Detroit Free Press* accused him of "plundering" by laying claim to chains and anchors that had broken loose from ships in the Detroit River. The newspaper asserted that Quigley collected these too quickly, without giving the owners a chance to recover them for themselves. Indignant, Quigley sought out the reporter and gave him a beating. When brought to court, Quigley admitted he punched the writer, but wanted to explain why he deserved it. The judge stopped him short, explaining that his guilty plea settled the case.

By the war's end, Quigley was captaining *Submarine No. 12*, the latest salvage vessel of the Western River Improvement and Wrecking Company, founded by James Eads. Eads had left the wrecking business before the war, but the enterprise continued to thrive. The Mississippi River campaigns of the war provided wreckers with a surplus of work. Quigley's *Submarine* was contracted to clear the Yazoo River in Mississippi of the wrecks of three federal gunboats, three Confederate gunboats, and twenty-four other military and commercial vessels that were hazards to navigation. It was thankless work, but Quigley did his part to heal the wounds of the great conflict.

When the war first broke out, Elliot P. Harrington, then residing in Iowa, heeded the call to arms and brought his family back to the Detroit area. There he volunteered his services and was hired as a contract salvager. His first assignment was to remove obstructions left by Confederate forces near Hampton Roads at Norfolk, Virginia. Later that year, he was put in command of the wrecking steamer *Dirigo* and dispatched to New Bern, North Carolina, to remove dozens of ships intentionally sunk to obstruct the port. From there, Harrington engaged in several wreck removals up and down the coast of South Carolina and Georgia, and assisted in offloading the heavy artillery guns (including the famed cannon, "Swamp Angel") that were used in the bombardment of Charleston, South Carolina, in late 1863. After hostilities ended, Harrington was still busily employed in the cleanup. In the summer of 1866 he was involved in raising hulks from the James River near Richmond, Virginia.

When the siege of Charleston was stalled in early 1863, Harrington wrote to Rear Admiral Samuel Francis Du Pont with an offer to build a submarine vessel; the experience he mentions could only have come from assisting Lodner Phillips with the *Marine Cigar* on Lake Erie. The full text of his letter (earlier quoted partially):

Steamer *Dirigo* May 11 1863

Sir: I am aware that many plans have been suggested for overcoming the obstructions in Charleston Harbor and that probably you may lack confidence in any new one. Still, I am anxious to present to you one of my own, which I have entire confidence in, which I trust you would consider entitled to careful consideration, if I should succeed in conveying my ideas plainly.

I have had twelve years' experience as a diver, have had enough to do in that line to render me familiar with all underwater operations. I raised the American Express safe from the wreck of the steamer *Atlantic*, in Lake Erie, at a depth of 170 feet of water, and have succeeded in several other undertakings of the kind which had been abandoned and declared impracticable by others. Among experiences was one with a submerged small propeller, driven by hand power, capable of being supplied with air by means independent of all outside help. With it I can make 1 1/2 miles per hour at a depth of 80 feet or less, and could conduct operations outside of it at any given depth with success. From my former experience with that craft, and my acquaintance with the whole subject, I am satisfied that I can construct a small propeller with which aided by from four to five men, I can without help from others and without being observed by the enemy, follow the channel at Charleston, cut the wires of torpedoes, cut any cables, or network of chains, saw off any piling, or overcome any other impediments likely to be met with. While doing so a telegraphic operation may be kept up with any monitor that may be detailed for that purpose and lying at a distance so that my own movements can be regulated or made known at any time.

I am aware that it looks like a hazardous and doubtful undertaking, yet after much experience I have such faith in it that I should be very glad to lay the details before you and leave it for your consideration; some difficulties that would at first strike an outsider as insuperable I am confident can be overcome.

Having so much faith in it, I would respectfully ask
of you the favor to grant me a personal interview and allow
me to detail my plans. Very respectfully, your obedient
servant, E P Harrington

There is no record of Du Pont's response to Harrington's offer, but
Du Pont had other parties with more impressive credentials approach-
ing him with the same concept. One was Lodner Phillips himself. By
1861, Phillips was residing in New York City, but had no working submarine models after the loss of the *Marine Cigar*. Phillips must have
been disgusted by the news of a "rebel torpedo boat," the *Hunley*, successfully attacking and sinking the USS *Housatonic*, in February, 1864.
In June 1864, he and a partner, Frederick M. Peck, presented plans for
the construction of three submarines to Secretary of the Navy Gideon
Welles. One was similar in dimensions to the *Marine Cigar*: 40 feet long,
capable of carrying five men underwater for twenty-four hours using
compressed air. It could be used to attach mines to enemy vessels or to
break them apart from detonations underneath. A second design was
for a submarine 120 feet long, capable of taking twenty men underwater
for five days. This model was armed with shell rockets and guns capable
of firing underwater. The third plan was for a submarine 200 feet in
length to patrol the seas and employ heavy guns.

Secretary Welles referred the proposal to the Permanent Commission of the Navy Department, which recommended appropriating
funds to build the smaller model, the one similar to the *Marine Cigar*.
The Commission's recommendation is as far as any effort by Phillips
went. There is no evidence that money was obtained or that a model
was built. It is likely that the Navy deferred any decision about Phillips's
designs or Harrington's offer because it had already approved construction of a submarine design invented by Brutus de Villeroi, who had
been experimenting with underwater vessels in France since the 1830s.
Villeroi's craft, the *Alligator*, became the first official submarine of the

United States Navy. Deployment of the *Alligator* had been promoted by a senior naval officer, none other than Samuel Francis Du Pont, the man to who Elliot Harrington had sent his letter. The *Alligator* had problems that were exposed during testing at the Washington Navy Yard, where it was decided to replace its awkward oars with a hand-cranked propeller. It was then towed toward the James River operations in March 1862, but a gale forced it to be scuttled near Cape Hatteras. The time and expense spent on the *Alligator* dampened the Navy's enthusiasm for submarine vessels.

Both Confederate and Union forces were more successful in adopting underwater explosives, collectively known at the time as "torpedoes." Any hesitation about the barbaric nature of unseen weaponry, or distinctions between offensive and defensive use, were overcome by the great imbalance between the opposing navies and the strategic and economic value of specific ports and waterways. Benjamin Maillefert was recruited to lead the Union's efforts to clear obstructions in the James River leading to Richmond, Virginia, and to blow up locks at South Mills, North Carolina. Maillefert, with access to more sophisticated electrical technology than his Confederate counterparts, was able to develop detonation timers attached to his charges. Even so, the Confederacy made more widespread use of torpedoes, and had enough success to insure that underwater explosives would remain a standard part of the naval arsenal.

Perhaps the most significant contribution made by a marine engineer to the Civil War efforts came from Mississippi wrecking magnate James Eads. When the war broke out, Eads offered Secretary of the Navy Gideon Welles the use of *Submarine No. 7* for Union operations on the Mississippi. Eads knew that his wrecking steamer's shallow draft and multiple hull compartments would have advantages over other river vessels. Intrigued, the Navy consulted with Eads to design a new type of river gunboat that would include heavy armor, good speed, and an array of cannons able to shoot from front, back and sides. The resulting design was the City class ironclad, similar in appearance to the well-known Confederate ironclad, the CSS *Virginia* (rebuilt from the previously scuttled USS *Merrimack*). Seven of the gunboats were built and formed the core of the Western Gunboat Flotilla, later known as the Mississippi River Squadron. Eads's gunboats presented an imposing sight to opposing vessels, but were mainly used against shore forts and defenses. They were first employed to capture Fort Henry. Later, they were instrumental in the taking of Fort Donelson, Island Number 10, and Fort Pillow. Ultimately, they were used to secure the river flank of Vicksburg while the Union Army had the city under siege. When Vicksburg fell, the Union won control of Mississippi and severely limited the ability of the Confederacy to adequately supply its western campaign.

As for John Green, he appears to have spent the Civil War years battling his internal demons. His only documented activity was one lecture at Milwaukee's Academy of Music in July, 1863. By war's end, Green's reliance on public sympathy for his disability would have been overwhelmed by the thousands of returning maimed war veterans, who also now depended on pensions and charity for their survival. From 1860 forward, Green boarded with his mother and sister Lucy in rooms on Seneca Street in Buffalo. In 1862, his sister Mary and brother-in-law Peter Tremble also lived in the same household. Tremble listed his vocation as "submarine diver," so it is possible that John Green still had a part in supervising other divers. However, the Buffalo newspapers did not make note of any of his activities.

Chapter Fifteen

≈

Ends (1866–1879)

After the war, diver Elliot Harrington did not return his life as a carriage maker in Iowa. It was probably a combination of two considerations that caused him to instead return to the Detroit area: first, his wife had been reunited with her family in Ypsilanti during the war years, and was reluctant to leave them again; second, Harrington's interest in diving had been rekindled by his recent service. After his return to Michigan, in October, 1865, Harrington chartered a ship to salvage the schooner *Willard Johnson*, which had been wrecked near Port Austin on Lake Huron.

However, once the federal government assessed the obstructions to navigation left by the war, Harrington was called east once again to help remove ships and debris from the James River near Richmond and Norfolk. At Norfolk, he worked on raising the old USS *Delaware*, which had been decommissioned long before the war began, but was scuttled to prevent its capture and use by the Confederacy. These contracts kept Harrington busy from the summer of 1866 through early 1867.

In late spring, 1867, at age forty-three, Harrington was caught up in the postwar mania for sports. It was during these years, before the savagery of boxing was constrained by accepted rules, that "collar-and-elbow" wrestling came to prominence as the country's premiere martial pastime. Derived from Irish martial arts and also known as "square-hold" wrestling, collar-and-elbow combatants used separate standing moves and, once one fighter was taken down, an array of ground techniques. Impromptu contests were made by challenge, with bettors setting aside a cash purse for the winner—and wagering many times that amount in side bets. There were no weight divisions, since skilled smaller men could often prevail against brute force. Harrington, at 174 pounds and (like most divers) several inches under six feet, accepted a wrestling match against thirty-seven-year-old Harrison Berdan, measuring six foot three inches and 209 pounds.

On the afternoon of the best-of-three round contest, all the betting was on Berdan. The grappling commenced with a series of feints and blocks, but Harrington used a "dexterous trip" to take down the larger man. Trips and kicks (without shoes) were legal, essential moves. The gamblers considered Harrington's initial success a fluke; the money remained on Berdan. After the between-round rest, the second round began. In thirty seconds Harrington was able to bring Berdan down again, and won the match. The account of Harrington's win was printed in newspapers around the Great Lakes.[1]

The fame only brought Harrington another challenge, this time for the unofficial Michigan "state championship." The *Detroit Free Press* reported from the scene in late April, 1867: "He [Harrington] stoops slightly, which gives him the appearance of being shorter than he really is, and steps heavily and deliberately, evincing no disposition to hurry; but all that is suddenly changed when he goes into action. His opponent, [Jacob] Martin, is only thirty-six years old, stands six feet in his stockings and weighs two hundred and fifteen pounds. Broad shouldered and muscular, he looks a perfect Hercules, and when he stepped into the ring the friends of Harrington began to lose confidence in his ability to cope successfully with his gigantic opponent." Harrington's stoop may have indicated that he, too, suffered residual effects of decompression illness, though not as severely as John Green. Against Martin, Harrington was unable to use the same tactics that had worked on the top-heavy physique of Berdan. Martin proved as quick with his moves as Harrington, and was able to toss the smaller man on his back twice in succession as quickly as Harrington had dispatched Berdan.[2]

Putting aside his wrestling sideline once the shipping season opened in 1867, Harrington returned to the water. In mid-June he was hired for a small job: recovering a brass cannon that had fallen overboard into the Detroit River from a Canadian gunboat, the *Prince Alfred*. He descended about thirty-five feet to the river bottom near Windsor, Ontario, tended by his assistants. A sharp report was heard from the support boat—Harrington's air intake hose had burst. Down below, he pulled on the signal line to be raised. Although the air intake had a valve that would prevent water getting in, the loss of pressurized air left the diver with few options. If he continued to expel exhausted air into the water, his suit would lose pressure and he would feel diver's squeeze, though at thirty-five feet it would not be extreme. If he shut his exhaust off, he would maintain some internal air pressure, but if he tried to hold his breath, as he was raised the air would expand and might cause a rupture of lung tissue.

With difficulty, Harrington was hauled onto the ship. The *Detroit Free Press* produced an account from eyewitnesses:

> The head-piece was instantly wrenched off, but what an awful spectacle greeted those who looked upon the sufferer. Blood was oozing from every part of his body from waist up, and gushed from his eyes, ears, and nostrils, while he was puffed and bloated beyond recognition. Both eyes were frightfully swollen, and his neck looked as if he had been choked severely, while with every movement blood gushed from his throat, his body and face meanwhile rapidly turning black. . . . Dr. Lewis was on the spot in a very few minutes, and several humane gentlemen went vigorously at work to restore circulation by rubbing him with brandy. His body and head continued to swell, and, as the

blood slowly dripped at his feet, many turned away, sick and faint at the ghastly sight.[3]

For hours it appeared that Harrington would suffer the same fate as divers William McDonnell and John Tope. However, later that evening, he was able to converse not only with his family, but with a *Detroit Free News* reporter. He said that he never lost consciousness, and that when the hose burst he immediately felt an immense pressure around the waist, and which almost squeezed his eyes out of their sockets.[4]

Within a short time, Harrington was back not only wreck diving, but wrestling. On September 19, 1867, he met William Payne of Ohio in a collar-and-elbow match at Merrill Hall in Detroit. In this case, the purse was $500; and the contest was best of nine rounds. Once again, Harrington grappled against a much younger and larger man, and lost two of the first three rounds, but then prevailed in all of the next four rounds. At the conclusion of the match, the *Detroit Free Press* reported that "Harrington holds himself in readiness to try his skill with any living man, win or lose." However, there is no record of any further matches taking place until two-and-a-half years later.[5]

Harrington's wrestling career came to a close in March, 1870, when he was nearly forty-six years old. An "International Wrestling Tourney" was held in Detroit, with America's best grapplers facing off against each other and fighters from Canada and Great Britain. There was no formally recognized wrestling championship in America yet, but this tournament was the first to award a belt as a prize. Harrington's first opponent was John Smith of Ontario, Canada. As the two stood face to face, there was a dispute over the position of the opening stance. The umpires ruled that Smith's positioning was not legal, but Harrington agreed to let the match continue. Harrington was thrown twice by Smith, but the umpires ruled that the result did not count.

Smith continued on to face the most feared wrestler in America, Col. James H. McLaughlin. McLaughlin was a large, intimidating figure, over six feet tall and weighing over 230 pounds. He had earned a reputation as a vicious, quick-tempered brawler in numerous army-camp contests during the Civil War, and by 1870 was already known to have permanently crippled several of his opponents. Some wrestling historians have called McLaughlin a psychopath. When McLaughlin met Smith, he quickly threw him to the ground three times in succession, but each time the umpires invalidated the moves. McLaughlin became enraged.

When the two men resumed their bout before the audience of Detroit's Young Men's Hall, McLaughlin literally lifted Smith up off his feet and threw him off the elevated stage. As a result, Smith fractured his skull and was unable to continue. The next day, he was found wandering aimlessly down the city streets, babbling nonsense. He was sent to an insane asylum and died shortly afterward. Later, when McLaughlin was in a more temperate mood, he sent some money to Smith's family. As a half-apology, McLaughlin offered, "I forgot I was so wicked strong."[6]

When it was Harrington's turn to face McLaughlin in Detroit, no one in the audience believed he had "a ghost of a chance."[7] This time, they were right—Harrington was tossed down twice in succession, but escaped being maimed. McLaughlin was declared the tourney's winner, and went on to dominate the wresting scene for much of the 1870s. Harrington judged it time to retire his wrestling laurels, and never again appeared in a public match.

Throughout the 1870s, Harrington gave several diving exhibitions in Detroit. On one occasion, he walked on the bottom of the river from Detroit to Windsor, and another time from Belle Isle to the end of Woodward Avenue in Detroit. His routine was much like that of George W. Taylor, thirty years earlier: a lecture on the history of diving apparatus, setting off a torpedo, firing a gun underwater, etc. He was billed as "Prof. E. P. Harrington, the World Renowned Diver."[8]

In March, 1875, Harrington led an expedition of Detroit divers to Hampton Roads, Virginia, in search of another safe full of riches. The company included his brother-in-law James Phillips and William Newton, one of the partners in raising the safe of the *Atlantic*. The team sought the strongbox of the sloop of war USS *Cumberland*, sunk by the Confederate ironclad CSS *Virginia* in March of 1862, which was rumored to contain $100,000. The wreck lay in about seventy-six feet of water in the James River. However, Harrington discovered that it had been worked over by many wreckers over the years, and that they had blasted the hulk with explosives. The safe, if not destroyed, might have been thrown some distance from the remaining timbers. After five weeks, they abandoned their efforts.[9]

Harrington began to cut back on his diving in the late 1870s, instead devoting his attention to machine-shop work. In 1876, he patented a railroad car coupling device that was safer than those in common use, but was not able to find any licensees. Harrington's next invention was quite different. The *Fort Wayne Gazette* printed this well-written article on Harrington in their September 1, 1879 edition:

A Buoyant Bicycle Propelled on the Water.

"Hey Mike! Come over here."

"What d'ye went?"

"Come over here and see the greatest sight ye ever saw. Here's a man ridin' a velocipede on the water."

The speaker was a man who stood in Hall's brick yards in Springwells [Detroit], hallooing to a friend some distance away. When the friend came over he saw, surely enough, a man on what appeared to be a bicycle propelling himself along over the water in the large pond in the brick yard.

Elliot P. Harrington is a submarine diver, well-known throughout the city. He often gives exhibitions. . . . One evening after one of the exhibitions at Bay City he was sitting on the wharf meditatively when a young boy rattled across the bridge on a velocipede.

"Why can't I make a thing like that run on water?" soliloquized Harrington. . . . the more he thought about the matter, the more he thought it possible, and he came home to Detroit and went to work. That was four weeks ago. Mr.

Harrington finished his machine which may be called an aquatic velocipede, in the strictest secrecy, and on Monday evening made a thorough and satisfactory trial of it on the large pond in Hall's brick yard.

The machine is in the form of a bicycle and is propelled with either the feet or hands or both. The wheel is of galvanized iron, hollow and air-tight and of great buoyancy. The "tire" consists of two flanges, six or seven inches apart; inside are the buckets or paddles which propel the wheel. The wheel is four feet in diameter and has pedals for the feet and cranks for the hands. In the relative position of the small wheel of a bicycle are two wooden fishes [pontoons] to which is attached a lever. They act as a rudder and also help to support the weight of the rider.

"How fast can you run it?" asked a reporter of the *Free Press.*

"Well, I didn't build this for speed. It is simply constructed to show the idea; however, I am going to Belle Isle some day this week and in the evening I shall run it down the river. I will give you signals by shooting sky-rockets into the air. Of course, I expect to improve the thing considerably yet."

"Do you expect to be able to travel long distances on the water with the aquatic velocipede?"

"Yes sir. I am going to Cleveland on it someday, and would like to have you ride behind me."

A reporter pleaded a predilection to seasickness.[10]

Just days after this experiment, Harrington was felled by a stroke. He returned to Chautauqua County, New York, thinking the country air breezes coming off Lake Erie might help him recuperate. He died there in the last days of October, 1879, at age fifty-five.

≈

Martin Quigley, the former senior partner of both John B. Green and Elliot P. Harrington, spent his entire later life as a wrecker and diver. Quigley had separated from his wife, the former Lucy Barnes, in the late 1850s. After relocating to St. Louis, Quigley married a young Irish woman named Ellen, thirty-one years his junior. Their only child, Ida Maud Quigley, was born in 1874, when Martin was sixty-three years old. Quigley conducted his work primarily on the Mississippi from the wharfs of St. Louis, where he probably spent as much time in diving bells as in submarine armor.[11] However, he also ranged up the Ohio to Cincinnati, and in 1875 returned to the Detroit area for wrecking contracts in Lake Huron.

In February, 1877, Quigley was recommended for an engagement at the newly opened New York Aquarium for what was billed as "Startling Subaqueous Performances." Inside a giant aquarium, "Captain Quigley, the Wonderful Submarine Diver, performs all kinds of manual labor, carpenter work, joiner work, etc., illustrating the manner of using the diver's armor." Sharing the tank with Quigley was nineteen-year-old Vivienne Lubin (real name Frances M. Watt), the "Water Nymph."[12] Lubin went on to star as the "water queen" with several circuses. For the gruff old diver Quigley, who had spent much of his career working alone in cold, muddy waters, the Aquarium act must have been a delight. Nothing in Quigley's public life suggested the villainous character "Perglich" that John B. Green implied. After his engagement at the New York Aquarium, Quigley spent his last years working for the government as a night watchman on the St. Louis piers. He died in that city in 1881 at age seventy.

~

Charles B. Pratt, the diver protégé of George W. Taylor, continued until 1866 his sporadic efforts to seek Taylor's last sought-after treasure, the payroll chest of the HMS *Hussar*. During the years he worked on the *Hussar*, Pratt spent the majority of his time in Worcester, Massachusetts, where he had settled in 1847. He continued diving infrequently until 1871. In 1876 he was elected mayor of Worcester, and subsequently reelected two more times. For many years, a helmet owned by Pratt, as well as artifacts recovered from the *Hussar*, were displayed in the historical rooms of the city of Worcester. Pratt, a pillar of the Worcester society, died in 1898.

～

James Aldrich Whipple, the most famous American diver prior to John B. Green, split his time in the last half of the 1850s between engineering work in Boston and wreck diving in the Caribbean. While in Boston, he applied for several patents: for improved pumps, a rotary steam engine, an improved steam boiler, a method for driving piles, and even a cigarette-rolling device. In 1860, he traveled to England and on to the Middle East to negotiate harbor construction contracts. He fell ill during this journey and returned to Boston to recuperate. Whipple died on August 9, 1861, his promising career cut short at the age of thirty-five.

～

Albert D. Bishop continued his marine engineering business in Brooklyn until his death in 1882. He bid for several contracts to raise ships, and proposed raising the CSS *Alabama*, the most successful Confederate raider of the Civil War, using a specially-built million-dollar floating derrick.[13] The government declined his offer. Both Bishop and Benjamin Maillefert bid for continued work on removing more of Hell Gate obstructions, Frying Pan Reef and Pot Rock. Maillefert received the contract, not Bishop. Bishop's Patent Derrick continued to receive attention, and was depicted on the cover of *Scientific American* in February, 1880, two years before his death.

～

Benjamin Maillefert spent most of his postwar years under contract with the United States government to remove wrecks resulting from the great conflict. His efforts were concentrated around Charleston harbor, and he now employed the new explosive, nitroglycerin. One of Maillefert's contracts called for the raising of the Confederate submarine *Hunley*, but he failed to locate it. Since Maillefert had been blasting the remains of the ship sunk by the *Hunley*, the USS *Housatonic*, many observers surmised that he had inadvertently destroyed the submarine. Despite years of working with explosives, other than the one serious accident at Hell Gate in 1852, Maillefert was never injured from blasting. However, he did nearly die in a famous fire that destroyed Richmond's Spotswood Hotel on Christmas Day in 1870. Maillefert had a productive life until succumbing to natural causes in 1884.

≈

After a last attempt to interest the federal government in his submarine designs in 1864, Lodner Phillips returned to more lucrative pursuits. In the late 1860s, he was living in New York City and experimenting with the new miracle material, plastics. In 1866 he patented a device for making buttons from celluloid plastics, one of the earliest practical applications for plastic. This might have proved to be a profitable direction for Phillips to channel his inventive genius, had he not been struck down by tuberculosis. Phillips died in New York in 1869 at the age of forty-three.

≈

John E. Gowen, after falling short of his bid to be a guano millionaire, agreed to a contract with the Russian government to clear the harbor at Sebastopol, the Crimean capital. During the Crimean War of 1854–1855, Sebastopol was put under siege by land and water by the allied forces of Britain, France, and the Ottoman Empire. To prevent attack from the sea, the Russians removed the ships' cannons and then deliberately scuttled their entire fleet to block the harbor. At the end of the war, Tsar Alexander II moved to reopen Sevastopol. The Russians had little experience with the newest wrecking techniques, and were faced with removing over one hundred hulks. The contract was awarded to John E. Gowen, representing the Philadelphia Submarine Mining Company, but the work area was infringed on by the one-time employer of divers James Whipple and John B. Green, the Boston Relief and Submarine Company. Gowen's was the more successful of the two wrecker companies, and gained world acclaim for his achievement. He went on to construct rail lines and steamer lines for the tsar, and over the next decades performed many salvage operations throughout Europe. He never resettled in the United States, dying in Paris in 1895.

≈

Henry Beaufort Sears took an example of his Nautilus diving bell to England, where it was praised. He employed it to maintain the Victoria docks in London for several years before turning his attention to agricultural devices. He made England his permanent residence and settled in the Liverpool area, where he filed several patents for grain- and silk-cleansing devices. His daughter Ella married Irvine Bulloch, a naval officer who served on the noted Confederate raider, the CSS *Alabama*. Bulloch chose not to return to the United States after the Civil War, which is perhaps one reason why Sears opted to stay in England.

Sears's movable diving bell work inspired another New York inventor, Van Buren Ryerson, to improve upon it by eliminating the need for the surface air supply. Ryerson's *Sub Marine Explorer* was patented in 1858 and put to practical use by blasting contractors Husted and Kroehl on Diamond Reef in Hell Gate. Julius Kroehl later improved on Ryerson's design to make a true submarine, also named the *Sub Marine Explorer*. In 1867, Kroehl took his submarine to the Pacific coast of Panama to engage in pearl fishing, but he died during the expedition. He probably succumbed to a recurrence of malaria he had first contracted during the Civil War.[14]

~

After the Civil War, James Eads's concentration turned from ironclad warships to the construction of a commercial bridge across the Mississippi at St. Louis. Other engineers deemed the project impossible and said that Eads's plans to rest bridge footing piers on the river bedrock were ill-advised. Eads believed the piers could be properly set by men working underwater in caissons, chamber-sized diving bells. The first caisson was sunk in late 1869. A few months later, the workers who had spent the longest periods underwater inside the caissons, breathing pressure-fed air, began to fall ill. Suspecting that their illness was related to the time they spent in the caisson, Eads ordered their shifts shortened. However, beginning in March, 1870, several of his workers collapsed and died. Eads brought in his family doctor, Alphonse Jaminet, to investigate the mysterious disease.

Jaminet suspected that breathing pressurized air caused the malady, but it varied so much from man to man and between veteran workers and new workers that he could not pinpoint the correlation between pressure and duration. He did, however, institute practices to increase the amount of time pressurizing and decompressing as men entered and left the caissons. Jaminet's research was continued six months later by Dr. Andrew H. Smith during the construction of the Brooklyn Bridge. Smith realized that pressurized air must be changing the absorption of oxygen in the blood. Like Jaminet, he instituted decompression measures for the caisson workers. Smith's identification of "Caisson Disease" was a giant step toward understanding that divers experienced the same condition, decompression sickness, which came to be known by the popular term "the bends."

~

Little is known about the activities of John Green in the years leading up to his death. His last recorded public lecture was in Milwaukee in July, 1863. Buffalo city directories indicate he boarded in rooms on Seneca Street with his mother and sister Lucy. For a period in the early 1860s, his other sister Mary and her husband Peter Tremble also lived in the same household. In 1865, Green's second wife, Grace A. Jennings, remarried back in Massachusetts. On her 1865 marriage record she used the last name Green, and indicated she had a former marriage, which suggests that she made no attempt to cover up her marriage to Green. However, no formal divorce documents have surfaced.

Green's disabilities never allowed him to find regular employment. During the 1860s, passages from his book were sometimes reprinted—in particular his descriptions of the glorious underwater scenes from Silver Bank—but sales from his work ceased to bring in any income. The cataclysm of war had diminished his personal feats, and public sympathy was stretched thin. In October, 1868, John B. Green swallowed a lethal dose of arsenic and slipped down into the unknown darkness for the last time.[15] His final resting place is unknown, but the most fitting spot would have been the site in Lake Erie where lies the wreck of the *City of Oswego*. Green had returned to the area many times during the 1850s. It is there that Green lost his soul, before he ever put on the submarine armor.

≈

If one views the progress of a technology as incremental steps of innovation, one building on another, then the American diving pioneers of the 1840s and 1850s were not landmark figures in that linear heritage. William Hannis Taylor's 1837 diving apparatus patent, compared to equipment that the Deane Brothers and Augustus Siebe had already developed in England, was an inferior, if not impractical, design. No actual Taylor equipment, or even graphic depictions of their apparatus, are known to have survived to the present day, and contemporary descriptions of their equipment lack accurate detail. It can be assumed that, during the 1840s, George W. Taylor adopted features of English gear, and Wells and Gowen clearly copied those designs from their start. By 1849–1852, the period when Whipple and Robinson, followed by Wells and Gowen, made dives over the USS *Missouri*, American equipment was described in one boast as equal to (or better than) any European equipment. It is not known what aspects of the diving apparatus—helmet, hoses, suit, air pumps—were included in that comparison.

It is also not known which brass workers made helmets for the Taylors, Whipple, or Wells and Gowen. All four of what would later become the most notable North American helmet manufacturers of the nineteenth century were active in the 1850s: Andrew J. Morse and Alfred Hale in Boston, August Schrader in New York, and John Date in Montreal. Schrader's corporate history notes that he made his first helmets after seeing divers off of Battery Park in 1846, a clear reference to an exhibition given by Taylor divers at Castle Garden in October, 1846. However, lacking documentary evidence, it is hard to determine when each of these businesses produced their first diving equipment. James Whipple's family preserved his letters and some drawings, and later added an advertisement for Alfred Hale equipment that was printed long after Whipple's death. An auction of Whipple artifacts also included a business card of Augustus Siebe. These are the only clues that the Taylor and Lake Erie divers might have been using equipment produced from recognizable manufacturers.

The diving helmet that Charles B. Pratt's granddaughter gave to the Worcester Historical Museum was "rediscovered" during the preparation of this manuscript. It has been on public display for over 120 years, but until now had escaped notice as a potentially significant artifact of diving history. The helmet bears no manufacturer marks, and appears to be a close imitation of a 1850s Siebe design. Whether it represents equipment willed to Pratt by George W. Taylor, or—more likely—a helmet Pratt had made later in the 1860s, may never be determined with certainty. The Pratt helmet is in such good condition that it looks like it was never or barely used. It includes an odd configuration of the air intake directly over the exhaust on the exact rear of the helmet. It may be a design that Pratt found impractical.

If the diving apparatus used by these pioneers had inconclusive links to the later main thread of diving technology, so too did the diving vessels built by Lodner Phillips and Henry B. Sears. Sears's concept of self-propelled diving bells was superseded by practical submarines for some applications and by improved diving apparatus for others, leaving untethered diving bells without even a niche market. Suspended diving bells still had advantages in rough water and in deep-water exploration, and large submerged enclosures (caissons) even today remain as tools of underwater construction. Sears's Nautilus bells were briefly viewed as being at the forefront of underwater exploration, but are now considered a footnote.

Lodner Phillips's place in diving history has been diminished more by his own tendency toward secrecy than by any other factor. If Phillips had publicized his experiments with the *Marine Cigar* on Lake Erie, and with the tests of his atmospheric diving suit, then it could be strongly argued that he might have influenced later inventors. But there is no evidence that later submarine designers ever saw any of Phillips's vessels in person, although the legacy of his work was known to officers of the U.S. Navy's Newport Torpedo Station, founded in 1869. By that juncture, many practical submarine designs existed. Many different designs for atmospheric diving suits were also promoted during the late nineteenth century, with very mixed results. It is regrettable that more is not known of Phillips's experience with his suit model.

The lasting significance of these pioneers is not in the mechanical details of their technology, but in the daring they exhibited and the limits they tested. They demonstrated that underwater engineering could be a practical business, that seeking treasure troves could sometimes be as romantic as its practitioners imagined, and that the undersea environment held unexpected wonders and dangers. The decompression illness that struck down John Green added to the evidence that breathing air under pressure had physiological effects that increased with time at depth, and eventually led to identification of the condition and development of tables that could define safe diving. The race to the *Atlantic* captivated the American public for half a decade through newspapers, and thanks to Green's autobiography, exposed generations of readers to the realm of underwater exploration. As with many tales of heroes, the stories about their deeds may have had a greater impact than the specific historical results of their actions.

Afterword

~

Envoi (1871–1891)

Long after the daring Lake Erie dives by Green and Harrington in the 1850s, a less-than-heroic figure continued to orchestrate the most-publicized diving ventures of the 1870s and 1880s. Daniel D. Chapin, the indefatigable metal dowser who was one of the first fortune-hunters attracted by the *Erie* wreck in 1843, continued his quest for precious metals for five more decades. Although Chapin was a peripheral figure in the story of the Lake Erie divers, a review of his career emphasizes a theme prevalent throughout this account: the fine line between entrepreneurial spirit and pathological avarice. Chapin epitomized the latter to such a degree that he almost appears to be the origin of the crazed "old prospector" character who has appeared frequently in modern popular culture.

Immediately following his several failed attempts to salvage the *Erie* in 1843–1845, Chapin purchased Round Island in the Hudson River, north of Peekskill (now connected with Iona Island). In 1845, he was observed mineral prospecting on the island, but he spent more time watching an effort underway about a half-mile down the river. There, at Jones Point, a ship was anchored and engaged in building a cofferdam around an old wreck in the belief that it was Captain Kidd's last treasure sloop. Legends suggested that, sometime around 1700, the privateer-turned-pirate purposely scuttled his ship while being chased by a British man-of-war. According to newspaper accounts of this 1845 salvage effort, the treasure hunters had been led to the spot by a clairvoyant. Chapin knew all about the stories of Kidd's treasure and had spoken to old-time area residents. Based on their information, he believed that the cofferdam was being built in the wrong place; he believed that Kidd's ship rested just off Round Island, which he now owned.[1]

Chapin spent the next twenty years mineral prospecting in New Jersey and the Hudson Valley, but he appears to have earned more money through speculating in mineral-rights leases on parcels of land near ore-rich Franklin, New Jersey. In the 1870s, Chapin became obsessed with a legend dating back to the French and Indian War of a strongbox containing a payroll equivalent to $85,000, which was lost in Lake Champlain during the 1758 Battle of Carillon (Ticonderoga). There are two different versions of this "lost payroll" treasure. One version claimed that British General James Abercrombie buried his troops' payroll strongbox before launching his ill-advised frontal assault on the French-held fort. Daniel Chapin believed the alternate version, which posited that a relief ship carrying French troops, provisions, and a large payroll capsized near Garden Island in Lake Champlain on its way to the fort. His repeated attempts to find the French treasure were fruitless.[2]

Failure did not discourage Chapin. In October of 1880, he returned to Round Island on the Hudson River with an old whaling schooner named the *Mary D. Leach* and a company of divers from the International Submarine Mining Company of New Haven, Connecticut, led by salvage engineer Charles F. Pike. Chapin directed the captain to anchor at a spot found by use of his "mineral compass." His device had changed little in the past forty years. Pike and Chapin spent a few unproductive weeks at the site, though Chapin was convinced they would discover Captain Kidd's lost treasure of $100,000 in gold pieces and an equal amount in silver.[3] When the first search failed, he convinced the backers of the venture to follow him to another rumored Captain Kidd site at a cove near the Bay of Fundy, hundreds of miles from Round Island. After that, too, proved fruitless, they returned again to the Hudson, where further searching turned up nothing.[4]

While most outside observers easily dismissed the hunt for Captain Kidd's treasure ship as folly, Chapin almost immediately turned up at the center of one of the most prolonged and costly sunken treasure hunts in American history, the search for the trove aboard the wreck of the HMS *De Braak* in Delaware Bay. The *De Braak* sank in May, 1798, during a sudden squall while escorting trade ships bound for Philadelphia. Britain was at war with both Spain and France, and French privateers were harassing British merchant vessels, necessitating an escort by naval ships.

Before reaching Delaware Bay, the *De Braak* had captured a Spanish cutter, the *Don Francisco Xavier*, bound for Cadiz from South America. The Spanish ship carried a cargo of copper and cocoa, but nothing of greater value. In 1805, seven years after the *De Braak* wreck, a local Delaware River pilot named Gilbert McCracken plotted the coordinates of the wreck site and told his son a story about saving Spanish crewmen of the *Xavier* who had washed overboard from the *De Braak*, where they were being held. Over the decades, local lore about the fact that the *De Braak* sank while carrying Spanish prisoners grew; the fact that the Spanish ship it had captured, the *Xavier*, had survived the storm intact was forgotten. McCracken's son Henry, and then his grandson Samuel, weaved the local anecdotes into a grandiose myth that the *De Braak* carried a load of Spanish gold, captured not just from one ship, but from many.

In 1877, Samuel McCracken tried to interest some investors in a search for *De Braak*'s treasure, but was unable to locate any evidence of the vessel based on his grandfather's coordinates. He recalled that many years earlier an old prospector had come to Delaware Bay and used a special instrument to pinpoint wreck sites. Coincidentally—or perhaps not—he received a letter from Chapin offering to help locate the *De Braak* just at the juncture when some action was needed to convince backers to continue the search.

In 1880, on advice from Chapin, Samuel McCracken prevailed upon the same speculative salvage company that sought Captain Kidd's treasure, the International Submarine Mining Company. Charles F. Pike and his crew then spent several seasons—and $30,000—searching for the *De Braak* treasure.[5] The spot that they anchored over was a sand mound that had been located years earlier by Chapin, although Pike's team clearly believed Chapin's ability to locate objects was limited to things that some person in his presence knew about. Chapin could wondrously find coins in the pockets of a group of people, and objects hidden by someone whose face Chapin could study. Today, it is clear that Chapin's talent was the mentalist's trick of reading body language.

In 1886, the International Submarine Company ("Mining" dropped from its name) was near failure from their own investment in the search. The effort was saved by a new lead investor, Dr. Seth Pancoast, an eminent Philadelphia physician. Pancoast poured his own resources—along with those of other investors he had ensnared—into three more years of searching. At one desperate point in 1888, Pancoast tried to pinpoint the precious metals by bringing in a mineral dowser he had found, a man known only as "Cline." Salvage engineer Charles F. Pike was present when Pancoast and Cline rode out to dowse over the dive site. After his experience ten years earlier with Chapin, it is a mystery why Pike would agree to this charade—unless Pike was working with "Cline" to pull the wool over the eyes of Pancoast.

By the late 1880s, the *De Braak* treasure search had already gained notoriety as a grand folly, but that did not prevent many more expeditions from being launched over the next hundred years, none of which found any hoard of gold. The last effort, in 1984, collected artifacts typical of a British naval vessel, but the salvage was so inept and destructive that marine archaeologists were horrified; the incident motivated the passage of the 1988 Abandoned Shipwrecks Act, which passed ownership of wrecks to state governments. This law was the best result to ever come from any of Daniel Chapin's treasure-hunting efforts.

Chapin spent his last years in the 1880s in Erie, Pennsylvania, trying to interest investors in raising the *Vermillion*, a freighter with a cargo of copper ore lost in Lake Erie. Chapin inflated the weight and value of the cargo with such hyperbole that prospective investors checked the ship's registry and noted that it was not designed to carry anything close to the amounts Chapin was suggesting. Chapin lived out his last days in an Erie boarding house, destitute and estranged from his family, who refused to even claim his body when he died in 1891. His death notice in the newspaper was headlined, "Wore Himself Out."[6] The degree to which Chapin was responsible for encouraging others to launch fruitless treasure-hunting expeditions, including all the efforts to raise the *Erie* and the *Atlantic*, will never be known.

Notes

Chapter One: Submarine Armor (1820s–1840)

1. A long, detailed account of the case against W. H. Taylor appeared on successive days in the *Baltimore Patriot*, "Communication: The Privateer *Federal*," Mar 10–11, 1830.

2. Taylor's trip to Washington to make his complaint is noted in an untitled article in the *Providence Patriot*, Feb 11, 1829.

3. The diving apparatus inventor "William H. Taylor" was identified with the family and background of William Hannis Taylor (1806–1848) as late as 1911, but then awareness of this connection faded. It was reexposed in James P. Delgado. 2012. *Misadventures of a Civil War submarine: iron, guns, and pearls*. College Station: Texas A&M University Press. The connection is confirmed by items appearing in Taylor's obituary in the *Washington Union*, "Obituary," Jul 2, 1848.

4. For details of the exploitation of Caribbean pearl divers, see Molly A. Warsh. 2010. "Enslaved Pearl Divers in the Sixteenth Century Caribbean." *Slavery & Abolition* 31:3 (Sep): 345–62.

5. W. H. Taylor and James Narine. 1837. *A new and alluring source of enterprise in the treasures of the sea, and the means of gathering them*: [Twelve lines of quotations]. New York: J. Narine, printer, 11 Wall Street, corner of Broad.

6. "Sub Marine Armour," *The Spectator* (New Bern, NC), Oct 12, 1838, p. 3.

7. "Submarine Excursion," *Farmer's Cabinet* (Amherst, NH), Aug 25, 1837.

8. Ibid. See also: "Pearl Fishery," *New York Evening Post*, Jul 17, 1837.

9. The "large vat" was probably not a glass-sided aquarium tank; those would become routine diving exhibition venues many decades later. The event

at Niblo's Garden was noted in an untitled column in the *Philadelphia Inquirer*, Oct 27, 1837.

10. By late October, 1837, Taylor's demonstrations ceased to mention the pearl expedition. For example, "By the Express Mail," *Southern Patriot* (Charleston, SC), Oct 27, 1837, p. 2.

11. Classified advertisement, *New York Evening Post*, Jan 5, 1838.

12. The author strongly suspects that "George W. Taylor, born 1807 in New Jersey," obscured his real name, his place of birth, or both. Despite dozens of articles detailing his exploits, he let slip no mention of his family circumstances or background. He first appears as a signer to William H. Taylor's 1837 patent application. No family other than his wife is mentioned on his death, even though his estate was contested.

13. "Submarine Armour," *New Hampshire Patriot*, Jun 18, 1838; and "Submarine Armour," *New Bedford Mercury*, Jun 22, 1838.

14. "Submarine Armor," *Florida Herald*, Nov 27, 1838.

15. "Electro-Magnetism," *Florida Herald*, Nov 27, 1838.

16. "Correspondence of the Journal of Commerce," *Boston Liberator*, May 7, 1841.

17. "Sub-Marine Search," *Baltimore Sun*, Mar 3, 1840.

18. "The Lexington," *Vermont Phoenix*, Apr 4, 1840.

Chapter Two: An Awful Calamity (1841–1844)

1. Found in Otis H. Tiffany. 1881. *Gems for the fireside, comprising the most unique, touching, pithy, and beautiful literary treasures from the greatest minds in the realms of poetry and philosophy, wit and humor, statesmanship and religion*. Philadelphia: Hubbard Brothers.

2. One of the Goodyear brothers accompanied the Taylors on their late 1838 trip to Florida, and was with G. W. Taylor during his demonstrations of the armor at Charleston, SC, in March and April of 1839. See "Sub-Marine Armour," *Southern Patriot* (Charleston, SC), Mar 7, 1839; and "Submarine Descent," *Philadelphia Public Ledger*, Apr 4, 1839. In several articles during this period, the equipment was described as "Taylor & Goodyear's Submarine Armor."

3. "Sub-Marine Armor—The Experiment," *Cleveland Evening Herald*, May 9, 1842.

4. "The Erie," *Cleveland Evening Herald*, Jun 18, 1842.

5. "The Fourth," *Cleveland Evening Herald*, Jul 2, 1842.

6. "Wreck of the Erie Found," *Cleveland Evening Herald*, Sep 5, 1842.

7. "Wreck of the Erie," *Cleveland Evening Herald*, Nov 24, 1842.

8. "Steamboat Erie Sunk," *Elyria* (Ohio) *Courier*, Dec 7, 1842.

9. "Narrow Escape and Probable Loss of Life," *Baltimore Sun*, Jan 11, 1843.

10. "The Wreck of the Erie," *Schenectady Cabinet*, Apr 25, 1843; "Interesting Discovery," *The New World* (New York), Apr 29, 1843; "The Wreck of the Erie Found," *Southport* (Wisconsin) *Telegraph*, May 2, 1843.

11. "A Diving Bell," *Brooklyn Daily Eagle*, May 15, 1843.

12. "Wreck of the Erie," *Cleveland Herald*, Jul 27, 1843.

13. The pessimistic view appears in "Wreck of the 'Erie,'" *Cleveland Plain Dealer*, Aug 28, 1845.

14. "Wrecks Discovered," *Milwaukee Weekly Sentinel*, Mar 2, 1844.

15. "The Erie," *Fredonia Censor*, Sep 10, 1844.

16. The *Buffalo Commercial Advertiser's* negative assessment was reprinted in [Untitled], *New Orleans Times Picayune*, Sep 26, 1844.

17. "Wreck of the 'Erie,'" *Cleveland Plain Dealer*, Aug 28, 1845.

Chapter Three: End of the Taylors (1840s–1850)

1. "Correspondence of the Journal of Commerce," *Boston Liberator*, May 7, 1841.

2. W. H. Taylor, 1842. *Salvage of the treasure ship Le Télémaque: lost in the River Seine, near Quillebeuf, on 3d January 1790, supposed to contain from 30,000,000 to 80,000,000 francs*. Havre: F. Hue.

3. The most detailed description, including drawings, of Taylor's technique for raising the Telemaque can be found in "Quillebeuf," *Illustrated London News*, Dec 17, 1842, pp. 504–5.

4. "The Telemaque," *Yorkshire Gazette*, Dec 24, 1842, p. 3.

5. "The Telemaque," *Hampshire Advertiser*, May 25, 1844; [Untitled], *Wiltshire Independent*, Mar 2, 1843; "The Telemaque," *Bury and Norwich Post*, Jan 4, 1843.

6. "The *Telemaque* Travestied," *Berkshire Chronicle*, Apr 1, 1843.

7. [Untitled], *Exeter Flying Post*, Sep 18, 1845.

8. "Bankrupts," *London Morning Chronicle*, Oct 22, 1845.

9. The service by Taylor aboard the *Macedonian* and his last plans are mentioned in his death notice, evidently written by his wife, Maria Airey Taylor: "Obituary," *The Union* (Washington, DC), Jul 2, 1848.

10. "Castle Garden—Grand Sub-Marine Explosion," *New York Tribune,* Aug 15, 1843.

11. Ibid.

12. See Philip K. Lundeberg, 1974. *Samuel Colt's submarine battery: the secret and the enigma.* Washington, DC: Smithsonian Institution Press.

13. "Schooner Spitfire," *Troy Daily Whig,* Aug 7, 1845.

14. "Horrific Submarine Adventure," *Adams Sentinel* (Gettysburg, PA), Jan 5, 1846.

15. The *Boston Evening Transcript* (Apr 21, 1846) and *New York Telegraph* reported that Taylor was charged with raising the *Missouri.*

16. Taylor gave a public demonstration of his new invention in New York: "The Fair Positively Closes" [advertisement], *New York Tribune,* Oct 22, 1846.

17. "New York—The War—Favoritism—Tampico," *New York Daily Tribune,* Oct 26, 1846, p. 2; "The War &c.," *New York Herald,* Jul 23, 1847.

18. By March of 1848, Taylor could be found removing obstacles from the Potomac River. It was there that one of his divers, Francis J. Wood, died in a diving accident, one of the first apparatus diving fatalities in America. See "Singular and Sudden Death," *Adams Sentinel* (Gettysburg, PA), Apr 4, 1848.

19. "Experiment with 'Chain Camels,'" *Baltimore Sun,* Jul 20, 1848.

20. Letters, Edward Robinson to James A. Whipple, August 1849–March 1850, Box 1, Folders 1–25, James A. Whipple Papers, William L. Clements Library, University of Michigan, Ann Arbor.

21. Franklin P. Rice. 1899. *The Worcester of eighteen hundred and ninety-eight. Fifty years a city.* Worcester, MA: F. S. Blanchard.

22. Entry for Charles B. Pratt; Ancestry.com. 2009. *1850 United States Federal Census* [online database]. Provo, UT: Ancestry.com Operations. Images reproduced by FamilySearch.

23. The most complete obituary is found in "Capt. George W. Taylor," *Trenton State Gazette,* Jun 10, 1850.

24. Letters, Edward Robinson to James A. Whipple, April 1850–June 1851, Box 1, Folders 1–25, James A. Whipple Papers, William L. Clements Library, University of Michigan.

25. Some advertisements for the submarine armor from the early 1850s omitted Taylor's name entirely, instead referring to "Goodyear's Submarine Armor." See "Submarine Armor" [advertisement], *Daily Alta California,* Feb 27, 1850.

Chapter Four: The Marine Engineers (1840s–1852)

1. "Pot Rock and the Whirlpool," *New London Daily Chronicle*, Jun 25, 1851, p. 2.

2. The most complete account of Phillips's career is found in Harris and Bolinger, *Great Lakes' first submarine*.

3. Ibid., pp. 23–24.

4. The anecdote concerning Eads and the reluctant diver occurred in 1842, during the time when Eads's first wrecking boat was under construction. Eads's biographer Florence Dorsey indicated "a professional Great Lakes diver from Chicago," but her source material was the article "Sketch of James B. Eads," *Popular Science Monthly* 28, Feb 1886. That article uses the phrase, "a diver from the lakes." There were no divers in 1842 on the Great Lakes other than the touring George W. Taylor.

5. "Common Pleas," *New York Commercial Advertiser*, Oct 17, 1840.

6. "Notice" [advertisement], *New York Tribune*, Oct 1, 1850.

7. "Raising Vessels on Lake Erie," *Cleveland Plain Dealer*, Apr 28, 1851.

8. "The May Flower To Be Got Off," *Milwaukee Sentinel*, Jan 6, 1852.

9. "A Novel and Useful Work," *Charleston Courier*, Jul 31, 1852.

10. Whipple's career is profiled in "The San Pedro Alcantara," *Gleason's Pictorial and Drawing-Room Companion*, vol. VI, no. 25, Jun 24, 1854, p. 400. This article includes a portrait of Whipple and a depiction of diving operations on the wreck of the *San Pedro de Alcantara*. Whipple's suggestion of exhausting air directly into the water was an improvement already found in Siebe helmets, which is a further indication that the Taylor equipment followed more primitive designs.

11. The full story behind the awarding of the Missouri contracts can be found in Chuck Veit. 2012. *Raising Missouri: John Gowen and the salvage of the U.S. steam frigate Missouri, 1843–1852*. [Raleigh, NC]: Lulu.com.

12. "Recovery of Powers' Statue of Calhoun," *New York Daily Tribune*, Nov 1, 1850, p. 5.

13. "Recovery of Lost Treasure," *Sandusky Register*, Sep 13, 1849; "The Brig *Plumper*," *Boston Courier*, Sep 27, 1849.

14. "Notice to Californians" [advertisement], *Boston Courier*, Mar 21, 1850.

15. "Digging Gold," *Baltimore Sun*, Jan 26, 1850.

16. "A Wrecking Expedition," *Cleveland Plain Dealer*, Apr 30, 1850.

17. "Wreck of the Lexington," *Cleveland Herald*, Nov 25, 1850.

18. The details of the raising of the USS *Missouri* can be found in Chuck Veit, *Raising Missouri*. A biographical entry for Stebbins can be found in Clark Waggoner. 1888. *History of the city of Toledo and Lucas County, Ohio*. New York and Toledo: Munsell & Co. A misunderstanding concerning conflicting sources on Stebbins's first name may have been caused by an incorrect cemetery headstone. "Daniel" is the name given in the 1850 and 1870 census records.

20. [Untitled], *Buffalo Commercial Advertiser*, Jun 17, 1844.

21. A complete description of the newly launched steamer can be found in "New Steamer G. P. Griffith," *Buffalo Commercial Advertiser*, May 3, 1848.

22. A full account of the wreck of the *G. P. Griffith* can be found in J. E. Hopkins. 2011. *1850: Death on Erie: the saga of the G. P. Griffith*. Baltimore, MD: PublishAmerica. Additionally, these articles provided anecdotes: "The Wreck and the Dead," *Cleveland Herald*, Jun 20, 1850; "Further from the Wreck," *Sandusky Register*, Jun 22, 1850; "Burning of the *Griffith*," *Trenton State Gazette*, Jun 22, 1850; and "Incidents of the Burning of the *Griffith*," *Ohio Statesman*, Jun 24, 1850.

23. Clark Waggoner. 1888. *History of the city of Toledo and Lucas County, Ohio*. New York and Toledo: Munsell & Co.

Chapter Five: The City of Oswego (July 1852)

1. A summary of the development of Great Lakes water craft can be found in Charles Patrick Labadie. 1990. *Minnesota's Lake Superior shipwrecks (A.D. 1650–1945)*. Saint Paul: Minnesota Historical Society, State Historic Preservation Office.

2. "The New Propeller," *Buffalo Daily Courier*, Jun 15, 1852.

3. Green's books are J. B. Green. 1859. *Diving, or, submarine explorations: Being the life and adventures of J. B. Green*. Buffalo, NY: Published by the Author; and J. B. Green. 1859. *Diving with & without armor: Containing the submarine exploits of J. B. Green*. Buffalo, NY: Faxon's Steam Power Press.

4. "The Census," *Oneida Morning Herald*, August 7, 1850

5. Ancestry.com. 2004. "1850 United States Federal Census" [online database]. Provo, UT: Ancestry.com Operations. Entry for John Green, Oswego, New York.

6. Green. 1859. *Diving, or, submarine explorations*, p. 5.

7. Ibid., pp. 8–9.

8. Ibid., p. 9.

9. Ibid., pp. 10–11.

10. Ibid., p. 11.

11. Ibid., p. 11.

12. Green, *Diving with & without armor*, p. 18.

13. One account that labels Ann Green as John's wife is also from the paper closest to the scene of the disaster: "Collision and Loss of Life," *Cleveland Leader*, Jul 13, 1852. Another was from Green's adopted hometown and port of departure: "The Late Disaster," *Oswego Daily Journal*, Jul 16, 1852. No marriage records from New York have surfaced that match a John and Ann(e) Green; however, marriage records from the 1840s were mainly church records and are very incomplete. There is a record of John Green's 1854 marriage to Grace A. Jennings; that record includes an indication that it is Green's second marriage. One strong candidate for proof of Green's first marriage is the union of John Green, sailor, with Ann Brady, at the Anglican St. Paul's Mariner Church in Quebec City, Quebec, in 1849. The ages of the two match those in other sources, and it's plausible that Green would have been working on a ship that went up the St. Lawrence to Quebec City. See Ancestry.com. 2008. *Quebec, Vital and Church Records (Drouin Collection), 1621-1967* [online database]. Provo, UT: Ancestry.com Operations. Original data: Gabriel Drouin, comp. *Drouin Collection*. Montreal, Quebec, Canada: Institut Généalogique Drouin. The author sent a copy of Green's second marriage record to Great Lakes historian Jack Messmer, who spotted the indication of a first marriage.

14. The $7,000 figure is found in Green, *Diving with & without armor*, p. 17. The $3,000 figure is cited in "Further Particulars of the Late Catastrophe," *Oswego Times*, Jul 29, 1852; the $300 figure was printed in "Further Particulars of the Late Catastrophe," *Cleveland Leader*, Jul 15, 1852.

15. Green, *Diving with & without armor*, p. 19.

Chapter Six: Without Armor and With Armor (July 1852)

1. Green, *Diving with & without armor*, p. 20.

2. Green did not name any of the diving party members other than the double-crosser "Perglich." However, the July 2, 1852, edition of the *Sandusky Register* notes: "Stebbins, part owner and engineer of the ill-fated Griffith, is now at work with a sub-marine diver, to raise the hull of that vessel and obtain some of her freight."

3. In October, 1852, Philadelphia's Franklin Institute sponsored an exhibition at which the Howard and Ash submarine armor was displayed. An article on the exhibition mentions that the armor had been used on Lake Erie the past

summer on the wrecks of the *Caspian* and the *City of Oswego* and was currently being used by divers working for Benjamin Maillefert on the *Atlantic* wreck. See "Local Affairs," *Philadelphia Public Ledger*, Oct 23, 1852. The armor used to dive to the *City of Oswego* in early August, 1852, was mentioned by a Cleveland paper as being the same armor that had been used to raise the USS *Missouri*, that is, a Wells and Gowen design. See "Mr. J. P. Gay," *Cleveland Herald*, Aug 2, 1852. Taking these sources together, it appears that the submarine armor used by Green and others on Lake Erie in 1852 was a Wells and Gowen design manufactured by Howard and Ash.

4. Green, *Diving with & without armor*, p. 23. Note that Green's first edition of his autobiography did not mention any name, including "Perglich." The second edition embellishes the stealing episode and calls out the villain "Perglich."

5. Ancestry.com. 2006. *Public Member Trees* [online database]. Provo, UT: Ancestry.com Operations. Entry for Martin Quigley, 1811–1881.

6. "The Atlantic," *Buffalo Commercial Advertiser*, May 26, 1849.

Chapter Seven: Mr. Wells's Safe (August–October 1852)

1. "The Steamer Atlantic for Sale," *Pittston* (PA) *Gazette*, Sep 3, 1852; "The Steamboat Atlantic," *New Orleans Times Picayune*, Sep 17, 1852.

2. "The Atlantic's Wreck," *Albany Evening Journal*, Aug 26, 1852.

3. Ibid.

4. "Disaster at Hurl Gate," *Brooklyn Daily Eagle*, Mar 27, 1852; "The Hurl Gate Catastrophe," *Baltimore American and Daily Advertiser*, Mar 31, 1852.

5. Green, *Diving with & without armor*, p. 24.

6. Ibid., p. 25.

7. See Letters, Edward Robinson to James A. Whipple, August 1849–March 1850, Box 1, Folders 1–25, James A. Whipple Papers, William L. Clements Library, University of Michigan. In these letters, Whipple and Robinson mention how their equipment was rated better than that used by any other Gibraltar divers.

8. See Letters, Ephraim B. Grant to James A. Whipple, 1852–1853, Box 1, Folders 1–25, James A. Whipple Papers, William L. Clements Library, University of Michigan.

9. "Yankee Enterprise," *Madison* (IN) *Dollar Weekly Courier*, Nov 17, 1852.

10. Green, *Diving with & without armor*, p. 24.

11. Ibid., p. 25.

12. The ineffective lamp is mentioned in a letter to an editor sent a few years later by Henry Wells: "The Atlantic Steamer & the Diver," *Kingston* (ON) *Daily News*, Jul 15, 1856, p. 2.

13. The Buffalo paper's claim that the previous record dive was 126 feet was reprinted in "The Steamer Atlantic," *New Orleans Times Picayune*, Oct 3, 1852.

14. Green, *Diving with & without armor*, pp. 25–26.

15. [Untitled], *Detroit Daily Advertiser*, Nov 12, 1852, p. 2.

Chapter Eight: The Erie Jinx (1853)

1. "Submarine Operations," *Trenton State Gazette*, Dec 18, 1852.

2. These objections were pointed out in Hopkins, *1850: Death on Erie*

3. "Wreck of the Erie," *New York Daily Tribune*, Jan 10, 1853, p. 5.

4. Reprinted in "Mr. John Green, the Diver," *Kalamazoo Gazette*, Dec 24, 1852.

5. Reprinted in "The Great Northern Diver," *Sandusky Register*, Jan 15, 1853.

6. "Yesterday Mr. Green," *Cleveland Plain Dealer*, Apr 19, 1853; [Untitled], *Buffalo Daily Republic*, Apr 26, 1853.

7. Green describes the operations over the *Erie* with no mention of his partner, Martin Quigley, at all: Green, *Diving with & without armor*, pp. 26–27.

8. "Suffocated," *Albany Evening Journal*, Jul 27, 1853.

9. Diver's squeeze is a danger particular to hard-helmet diving in non-atmospheric suits. At its extreme, the dramatic effects have been subject to legend, and also to contemporary confirmation from the American television program, *Mythbusters*: "Mythbusters—Compressed Diver," YouTube video, 2:47, posted by "Funn," Nov 26, 2009, http://youtu.be/LEY3fN4N3D8 (accessed Oct 14, 2014).

10. Reprinted in "Something of an Undertaking," *Sandusky Register*, Aug 18, 1853.

11. "The Atlantic," *Milwaukee Daily Free Democrat*, Sep 12, 1853; [Untitled], *Detroit Free Press*, Sep 1, 1853.

12. Letters, Daniel Driscoll to James A. Whipple, 1853–1854, Box 1, Folders 1–25, James A. Whipple Papers, William L. Clements Library, University of Michigan.

13. Ibid.

14. "Local News," *Buffalo Daily Courier*, Sep 15, 1853.

15. [Untitled], *Buffalo Commercial Advertiser*, Sep 24, 1853; "Effects of the Gale," *Buffalo Daily Republic*, Sep 24, 1853.

16. "Serious Disaster on Lake Erie," *Albany Evening Journal*, Sep 26, 1853.

17. Green, *Diving with & without armor*, p. 28.

Chapter Nine: Harrington and the Diving Boat (October 1853–Spring 1854)

1. Ancestry.com. 2006. *Public Member Trees* [online database]. Provo, UT: Ancestry.com Operations. Entry for Elliot Perry Harrington, 1824–1879.

2. One paper printed news of the *City of Oswego* sinking and the Westfield fire under the same headline: "Propeller Oswego," *Boston Evening Transcript*, Jul 15, 1852.

3. Credit for catching the coincidence of Harrington's store burning down and the wreck of the *City of Oswego* goes to Great Lakes historian Jack Messmer. Mr. Messmer communicated this to the author in a 2013 email.

4. The particulars of Harrington's first salvage dive must be pieced together from the historical record of shipwrecks and misremembered anecdotes. For instance, an 1867 Detroit newspaper profile of Harrington states that he first dove in 1850 on the wreck of the *Princeton*, but that wreck occurred in 1854. A 1915 lecture prepared by F. A. Hall for the Chautauqua County Historical Society stated his first dive was the *Oneida* in 1852. An 1873–1874 Chautauqua County Gazetteer also indicates that Harrington first dove in 1852.

5. L. D. Phillips. 1852. Steering Submarine Vessels. US Patent 9,389 dated November 9, 1852.

6. Reprinted in "A Sub-Marine Propeller," *Buffalo Daily Republic*, Oct 15, 1853.

7. Ibid.

8. United States, and Richard Rush. 1894. *Official records of the Union and Confederate navies in the war of the rebellion*: ser. I, vols. 1–27, ser. II, vols. 1–3. Washington, DC: Government Printing Office. Harrington's letter appears in ser. 1, vol. 14, p. 187.

9. Edmund L. Zalinski, "Submarine Navigation," *Forum*, vol. 2 (Jan), 1887, p. 474–75.

10. "Harrington the Diver," *Marine Review*, Apr 15, 1897, p. 13.

11. A letter published in 1859 confirms that the atmospheric suit had been built. The letter's author was only indicated by initials: "N. W." He claims

to have built it "three or four years since," i.e., in 1855–1856. He also states that it was designed to descend to 500 feet (!). See "Submarine Navigation," *New York Evening Post*, Aug 16, 1959.

12. See Letters, Daniel Driscoll to James A. Whipple, 1853–1854, Box 1, Folders 1–25, James A. Whipple Papers, William L. Clements Library, University of Michigan.

13. "Submarine Wrecking," *Madison* (WI) *Dollar Weekly Courier*, Jun 14, 1854.

14. "Submarine Armor," *New Orleans Times Picayune*, Dec 14, 1854.

Chapter Ten: Boston Bliss (1854–July 1855)

1. Green's Boston visit is noted in [Untitled], *Boston Post*, Jan 25, 1854.

2. "The Wrecks of the Steamer Atlantic and Erie," *Buffalo Daily Republic*, Apr 25, 1854.

3. "Submarine Operations on the Wreck of the *Humboldt* and *Staffordshire*," *New York Evening Post*, Apr 19, 1854.

4. "Two Locomotives Found in the Lake!," *Daily Forest City* (Cleveland, OH) *Democrat*, Mar 3, 1854.

5. "The Steamer *Erie*," *Troy Daily Whig*, May 6, 1854.

6. Tope was no ordinary seaman. He was named as the "engineer" charged with locating the wreck of the *Staffordshire* at Cape Sable. His descendants knew him to be a "sea captain." After his death, his family emigrated from the United States to Australia, where his brother started a prosperous business. Oddly, John E. Gowen filled out a United States passport application for Tope, indicating he was born in Barnstable, Massachusetts. In reality, he was born in Devonshire, England. See Ancestry.com. 2006. *Public Member Trees* [online database]. Provo, UT: Ancestry.com Operations. Entry for John Tope, 1822–1854.

7. "Horrid Death of a Sub-Marine Diver," *Cleveland Daily Herald*, May 31, 1854; "The Late Death in Sub-Marine Armor," *Philadelphia Inquirer*, Jun 2, 1854.

8. Green, *Diving with & without armor*, pp. 29–31.

9. "Wreck of the Erie," *Watertown* (NY) *Chronicle*, Aug 30, 1854; "Wreck of the Erie," *Southern Weekly Post* (Raleigh, NC), Sep 2, 1854.

10. Ancestry.com. 2013. *Massachusetts, Marriage Records, 1840–1915* [online database]. Provo, UT: Ancestry.com Operations. Entry for John Green and Grace A. Jennings.

11. "Submarine Explosion," *Skaneateles Democrat*, Dec 1, 1854.

12. "Dissolution," *Boston Courier*, Jan 1, 1855.

13. Green, *Diving with & without armor*, pp. 31–32.

14. Boston Submarine and Wrecking Co. 1854. *Circular of the Boston Submarine and Wrecking Co., November, 1854.* Boston: Geo. C. Rand, printer, 3 Cornhill.

15. Green, *Diving with & without armor*, p. 33.

16. "Later from Turk's Island," *New York Evening Post*, Jul 20, 1855.

17. Green, *Diving, or, submarine explorations*, p. 23.

18. Green, *Diving with & without armor*, p. 38.

19. Ibid., p. 40.

Chapter Eleven: Race to the Atlantic
(August–December 1855)

1. "The Sea Serpent Astray," *Oneida Sachem*, Aug 4, 1855.

2. "Experiment Company," *Skaneateles Democrat*, Sep 12, 1855.

3. Green, *Diving with & without armor*, pp. 40–42.

4. "Diving for Gold," *New York Tribune*, Sep 6, 1855, p. 6.

5. Green, *Diving with & without armor*, p. 41.

6. Ibid., p. 42; also "The Treasure Chest of the Atlantic," *Brooklyn Daily Eagle*, Sep 6, 1855. A log book can be found today in the Port Dover Harbour Museum with Captain Patterson's entry. The museum is also filled with *Atlantic* artifacts recovered by diver Mike Fletcher.

7. "John Green the Diver," *Buffalo Daily Courier*, Sep 8, 1855.

8. "Recovering," *Rome (NY) Sentinel*, Nov 20, 1855.

9. Green, *Diving with & without armor*, pp. 42–43.

10. "Submarine Blasting," *Hunt's Merchants' Magazine and Commercial Review*, vol. 30, Jan–Jun 1854, pp. 191–96.

11. The contentious bidding process for removal of Diamond Reef is explained in James P. Delgado, *Misadventures*.

12. Harris and Bolinger, *Great Lakes' first submarine*, p. 30.

13. Lincoln Diamant. 1989. *Chaining the Hudson: The fight for the river in the American Revolution*. New York: Carol Pub. Group.

14. "Diving Bell Experiment," *Brooklyn Daily Eagle*, Nov 17, 1853.

15. "A Novel Enterprise," *Washington Daily Union*, Jan 3, 1855.

16. "To Submarine Operators" [advertisement], *New York Evening Post*, Jan 19, 1855.

17. "Submarine Exploration," *Portland Weekly Advertiser*, Apr 24, 1855; "From Our New York Correspondent," *Washington Daily Intelligencer*, Apr 20, 1855.

18. "Diving Apparatus," *Sacramento Daily Union*, Apr 2, 1856; "For Sale: The Barque Emily Banning" [advertisement], *Daily Alta California*, May 21, 1856.

19. All the details concerning Gowen's guano ventures can be found in Chuck Veit.2014. *The Yankee expedition to Sebastopol: John Gowen and the raising of the Russian Black Sea fleet, 1857–62*. [Raleigh, NC]: Lulu.com. The author is grateful that Mr. Veit shared this chapter of Gowen's story prior to its publication.

20. "Treasure of the Frigate Huzzar," *Milwaukee Weekly Wisconsin*, Jun 29, 1855; "Sub-Marine Diving," *New Orleans Times Picayune*, Jul 23, 1856.

21. "The Permanent Wharf at Pensacola Navy-Yard," *Washington Daily Union*, Mar 27, 1855; "Pensacola Harbor," *Wilmington (NC) Journal*, May 16, 1856.

22. "Extraordinary Operations of a Diver Under Water," *Kenosha (WI) Democrat*, Jun 9, 1854.

23. "Raising of the Propeller Princeton," *Buffalo Daily Republic*, Sep 7, 1855.

24. Ancestry.com. 2006. *Public Member Trees* [online database]. Provo, UT: Ancestry.com Operations. Entry for Elliot Perry Harrington, 1824–1879.

25. "Wreck of the Oregon," *The Democracy* (Buffalo, NY), Jun 8, 1855.

26. "The Lakelets in Michigan," *New York Tribune*, Jun 15, 1855; "A Curious Lake in Michigan," *Syracuse Daily Standard*, Jul 16, 1855.

27. "The Adventures of a Diver," *Detroit Free Press*, Jun 9, 1867.

28. Ibid.

29. "River Matters," *St. Paul Daily Pioneer*, Mar 11, 1856.

30. "The Adventures of a Diver," *Detroit Free Press*, Jun 9, 1867.

Chapter Twelve: The Safe Recovered (1856)

1. Ancestry.com. 2006. *Public Member Trees* [online database]. Provo, UT: Ancestry.com Operations. Entry for Theophilus Peter Tremble, 1831–1893. The anecdote about marrying the baby Mary Green is found in "Theophilus Peter Tremble," WeRelate, http://www.werelate.org/wiki/Person:Theophilus_Tremble_%281%29, accessed Sep 21, 2014.

2. Green, *Diving with & without armor*, p. 44.

3. M. Osborn. 1858. *Apparatus for Raising Sunken Vessels.* U.S. Patent 20,578; publication date Jun 15, 1858.

4. [Untitled], *Charleston (SC) Courier,* Jun 26, 1856.

5. "The Wreck of the Atlantic," *New York Tribune,* Jul 3, 1856; "The Raising of the Safe of the Atlantic," *Cleveland Herald,* Jul 8, 1856; "The True Story of the Diver and the Wreck of the Atlantic," *Savannah Daily Morning News,* Jul 18, 1856.

6. The *Detroit Advertiser*'s fanciful version was reprinted in "The Wreck of the Atlantic," *New York Tribune,* Jul 3, 1856.

7. "The Safe of the Atlantic," *New York Herald,* Jul 16, 1856.

8. Once again, Harrington's diving tender from the 1860s, T. E. Kinney, provides a point of interest regarding the issue of whether Green's efforts made the safe recovery easier for Green. Kinney, writing in 1897, offers a convoluted version of events, remembered from the story Harrington had told him in the 1860s: "[Green] reached the wreck and broke in one of the windows of the clerk's office. He also managed to put in a small anchor line, but upon coming up from the wreck after partial success, he was unable to proceed further and died soon afterwards from the effect of injuries which he sustained in the work. . . . [Harrington] upon reaching the hurricane deck, found Mr. Green's anchor line. He followed the line to the clerk's office, where he found the broken window and anchor inside." See "Harrington the Diver," *Marine Review,* Apr 15, 1897, p. 13.

From these muddied, conflicting accounts, this writer believes that Harrington found Green's line and followed it to the cabin, and that Harrington extracted the safe from the cabin, not Green.

9. "Kid, the Pirate," *Oswego Palladium,* Sep 11, 1856.

10. "An Interesting Relic," *Buffalo Courier,* Jun 26, 1884.

Chapter Fifteen: Ends (1866–1879)

1. "Wrestling Match Between Berdan and Harrington," *Milwaukee Sentinel,* Apr 22, 1867.

2. "A Wrestling Match," *Detroit Free Press,* Apr 29, 1867, p. 1.

3. "Frightful Accident," *Detroit Free Press,* Jun 20, 1867, p. 1.

4. "The Perils of Submarine Diving," *Albany Evening Journal,* Jun 24, 1867.

5. "The Wrestle Last Evening," *Detroit Free Press,* Sep 20, 1867.

6. For a summary of McLaughlin's fearsome career, see Tim Corvin. 2014. *Pioneers of professional wrestling: 1860–1899*. Bloomington, IN: Archway.

7. "International Wrestling Tourney," *Spirit of the Times*, Mar 15, 1870, p. 75.

8. "Submarine Exhibition" [advertisement], *Detroit Free Press*, Sep 11, 1874.

9. "The Cumberland Expedition," *Detroit Free Press*, Apr 24, 1875, p. 1.

10. "A Buoyant Bicycle Propelled on the Water," *Fort Wayne Gazette*, Sep 1, 1879.

11. Ancestry.com. 2006. *Public Member Trees* [online database]. Provo, UT: Ancestry.com Operations. Entry for Martin Quigley, 1811–1881.

12. "The Great New York Aquarium" [advertisement], *New York Herald*, Feb 11, 1877.

13. "The Alabama," *National Republican*, Mar 13, 1863.

14. The story of Kroehl's submarine and its remarkable recent rediscovery is told in Delgado, *Misadventures*.

15. "Suicide of John G. [sic] Green, the Diver," *Washington Daily National Intelligencer*, Oct 29, 1868.

Afterword: Envoi (1871–1891)

1. "Captain Kidd's Vessel," *Milwaukee Daily Sentinel*, Aug 4, 1845.

2. "Minor Topics," *Jersey City Evening Journal*, Nov 22, 1877.

3. As they toiled on the river, 600 feet away on Round Island itself, Chapin had leased quarry rights to the builders of the Brooklyn Bridge. Huge blocks were cut out and placed on barges, and then floated down the Hudson, where they were set in the bridge footings by John Augustus Roebling's caisson workers.

4. "Is it Capt. Kidd's Ship?" *Louisiana Capitolian*, Nov 3, 1880.

5. "Bay-Bottom Treasure," *Philadelphia Times*, Oct 23, 1881.

6. "Wore Himself Out," *Cleveland Leader*, Jul 1, 1891.

Bibliography

Bauer, K. Jack. 1974. *The Mexican War, 1846–1848.* New York: Macmillan.

Boston Submarine and Wrecking Co. 1854. *Circular of the Boston Submarine and Wrecking Co., November, 1854.* Boston: Geo. C. Rand, printer, 3 Cornhill.

———. 1856. *Report of the board of directors of the Boston Submarine and Wrecking Co., through their president, S. Benton Thompson, submitted to the stockholders at their annual meeting, March 12, 1856.* Boston: Press of the Franklin Printing House.

Corvin, Tim. 2014. *Pioneers of professional wrestling: 1860–1899.* Bloomington, IN: Archway.

De Kay, James T. 1995. *Chronicles of the frigate Macedonian, 1809–1922.* New York: Norton.

Delgado, James P. 2012. *Misadventures of a Civil War submarine: Iron, guns, and pearls.* College Station: Texas A&M University Press.

Diamant, Lincoln. 1989. *Chaining the Hudson: The fight for the river in the American Revolution.* New York: Carol Pub. Group.

Dorsey, Florence L. 1947. *Road to the sea: The story of James B. Eads and the Mississippi River.* New York: Rinehart.

Dunn, Pete J. 2002. *Mine Hill in Franklin, and Sterling Hill in Ogdensburg, Sussex County, New Jersey: Mining history, 1765–1900: Final report.* Alexandria, VA: P. J. Dunn.

Frew, David R. 2004. *Interrupted journey: The saga of the steamer Atlantic.* Erie, PA: Erie County Historical Society and Museums.

Goodman, Martin. 2008. *Suffer and survive: Gas attacks, miners' canaries, space-suits and the bends—the extreme life of J. S. Haldane.* London: Pocket Books.

Green, J. B. 1859. *Diving, or, submarine explorations: Being the life and adventures of J. B. Green.* Buffalo, NY: Published by the Author.

————. 1859. *Diving with & without armor: Containing the submarine exploits of J. B. Green.* Buffalo, NY: Faxon's Steam Power Press.

Harris, Patricia A. Gruse, and Mary Thomas Bolinger. 2010. *Great Lakes' first submarine: L.D. Phillips' "Fool Killer."* Michigan City, IN: Michigan City Historical Society.

Hopkins, J. E. 2011. *1850: Death on Erie: The saga of the G. P. Griffith.* Baltimore, MD: PublishAmerica.

Huhtamo, Erkki. 2013. *Illusions in motion: Media archaeology of the moving panorama and related spectacles.* Cambridge, MA: MIT Press.

Karras, Mead Smith. 2011. *Commodore Josiah Tattnall: From pirates to ironclads, half a century in the old navy.* Bloomington, IN: Authorhouse.

Labadie, Charles Patrick. 1990. *Minnesota's Lake Superior shipwrecks (A.D. 1650–1945).* Saint Paul: Minnesota Historical Society, State Historic Preservation Office.

Lundeberg, Philip K. 1974. *Samuel Colt's submarine battery: The secret and the enigma.* Washington, DC: Smithsonian Institution Press.

Mattson, Arthur S. 2009. *Water and ice: The tragic wrecks of the Bristol and the Mexico on the South Shore of Long Island.* Lynbrook, NY: Lynbrook Historical Books.

McInnis, Maurie Dee. 2005. *The politics of taste in antebellum Charleston.* Chapel Hill: University of North Carolina Press.

New York (State) Department of State. 1791. *Register of gold and silver mines.*

Oickle, Alvin F. 2011. *Disaster on Lake Erie: The 1841 wreck of the steamship Erie.* Charleston, SC: History Press.

Phillips, John L. 1998. *The bends: Compressed air in the history of science, diving, and engineering.* New Haven, CT: Yale University Press.

Powers, Dennis M. 2009. *Taking the sea: Perilous waters, sunken ships, and the true story of the legendary wrecker captains.* New York: AMACOM.

Runyan, Timothy J. 1987. *Ships, seafaring, and society: Essays in maritime history.* Detroit: Published for the Great Lakes Historical Society by Wayne State University Press.

Shomette, Donald G. 1993. *The hunt for HMS De Braak: Legend and legacy.* Durham, NC: Carolina Academic Press.

Swayze, David D. 1992. *Shipwreck! A comprehensive directory of over 3,700 shipwrecks on the Great Lakes: Includes the most dangerous spot on the lakes, largest freighters lost, the most dangerous decade, treasure ships.* Boyne City, MI: Harbor House Publishers.

Taylor, W. H. 1842. *Salvage of the treasure ship Le Télémaque: Lost in the River Seine, near Quillebeuf, on 3d January 1790, supposed to contain from 30,000,000 to 80,000,000 francs.* Havre: F. Hue.

Taylor, W. H., and James Narine. 1837. *A new and alluring source of enterprise in the treasures of the sea, and the means of gathering them:* [Twelve lines of quotations]. New-York: J. Narine, printer, 11 Wall Street, corner of Broad.

Taylor's Submarine Pearl Fishing Company, and William H. Taylor. 1848. *Taylor's Submarine Pearl Fishing Company: A new and alluring source of enterprise in the treasures of the sea, and the means of obtaining them.* New-York: Van Norden & King, stationers, no. 45 Wall-Street.

Varhola, Michael O. 2008. *Shipwrecks and lost treasures, Great Lakes: Legends and lore, pirates and more!* Guilford, CT: Globe Pequot.

Veit, Chuck. 2012. *Raising Missouri: John Gowen and the salvage of the U.S. steam frigate Missouri, 1843–1852.* [Raleigh, NC]: Lulu.com.

————. 2014. *The Yankee expedition to Sebastopol: John Gowen and the raising of the Russian Black Sea fleet, 1857–62.* [Raleigh, North Carolina]: Lulu.com.

Viele, John. 2001. *The Florida Keys.* Vol. 3: *The wreckers.* Sarasota, FL: Pineapple Press.

Wachter, Georgann, and Michael Wachter. 2000. *Erie wrecks east: A guide to shipwrecks of Eastern Lake Erie.* Avon Lake, OH: Corporate Impact.

————. 2001. *Erie wrecks west: A guide to shipwrecks of Western Lake Erie.* Avon Lake, OH: Corporate Impact.

Warsh, Molly A. 2010. "Enslaved Pearl Divers in the Sixteenth Century Caribbean." *Slavery & Abolition,* 31:3 (Sep): 345–62.

Wells, Jamin. 2012. " 'Plenty of Glory but No Dividends': Marine Salvage and the Lore of the Shore in Late Nineteenth-Century America." *Research Seminar Paper* 144 (Apr 19). The Center for the History of Business, Technology, and Society, Hagley Museum and Library.

Index

Abandoned Shipwrecks Act, 168
Alabama (Lake Erie steamer), 100
Alligator (submarine), 145
America (steamer), 63
American Express Company, 72, 76, 81, 88, 123, 125
Atlantic (steamer), 71–73, 75–76, 78–81, 86, 89, gallery 12, 93, 107–109, 119–125
atmospheric diving suits, gallery 9, 95, 180–181n11

Baltic (steamer), 88–89
Barber, Francis Morgan, Lt., 40
Battle of Lake Erie, 5
bends. *See* decompression sickness
Bishop, Albert D., 42–44, 81, 86–90, gallery 6, 111, 157
Bishop's Patent Derrick, 43, 48–49, 81, 86–90, gallery 6, 157
Boston Locomotive Works, 44
Boston Sub-Marine and Wrecking Company, 101
Boston Relief and Submarine Company, 102, 159
Bradstreet, Peter G., 89
Bristol (ship), 8
Brooklyn Bridge, 109, 160, 185n3

Bushnell, David, 37
Buzzard, HMS, 42

caisson disease. *See* decompression sickness
camels (lifting apparatus), 29–30, 48
Canton (brig), 28
Caspian (steamer), 68, 84
Captain Kidd's treasure, 125, 165–166
Castle Garden, New York City, 27
Chapin, Daniel D., 18–21, 124–125, 165–169
Chatsworth (ship), 48
City of Buffalo (steamer), 140
City of Oswego (propeller), 58–59, 62–66, 90, 129–130, 180n2–3
clairvoyant locators, 17, 19, 166
Coffin, Isaac, 97, 100–101
collar-and-elbow wrestling, 149–153
Colt, Samuel, 28
Columbia (steamer), 80
Cooley, Harrison R., 139–141
coopering, 23–24
Cumberland, USS, gallery 15, 153–154

Date, John, 162

De Braak, HMS, 167–168
De Villeroi, Brutus, 145
Deane, Charles and John, 6
decompression sickness, 108–109,
    150, 160
DeKay, George (Commodore), 4, 26
Delaware, USS, 149
DeWitt Clinton (ship), 13
Dirigo (steamer), 143
diver's squeeze, 85, 98–99, 151–152,
    179n9
diving bells, 6, 19–20, 35–36, 41, 78,
    163
dowsing, mineral. *See* mineral
    compass
Driscoll, Daniel, 88, 90, 95, 111
Du Pont, Samuel Francis, Rear
    Admiral, 94, 143–146

Eads, James, 40–42, gallery 4, 116,
    142, 146–147, 160, 175n4
Eagle (brig), 42
electric motors, 23
Elizabeth (ship), 45
Emily Banning (barque), 112–113
Empire (steamer), 89
Erie Canal, 11, 50, 57–58, 62, 141
Erie (steamer), 11–16, 19–21, 69, 71,
    84–85, 89, gallery 1, gallery 13,
    97–100
Erie (steamer, a.k.a. *Little Erie*). See
    Little Erie
Erie, USS, 3–5, 11
explosives, underwater, 33–35,
    76–77, gallery 3

Federal (schooner), 3–5, 11
Fletcher (schooner), 120

Fool Killer (submarine), 38
Foreman, Edgar W., 35–36, 75
Forest Queen (steamer), 117
Fox (steamer), 78–79
Fox, Watson A., 43
Francis Metallic Life-Boat, 76
Fuller, Luther, 12–14
Fuller, Margaret, 45
Fulton, Robert, 37

G. P. Griffith (steamer), 49–53,
    67–69, 83–84, gallery 6, 176n22
Gardner, Charles O., 77, 91, 120
gold prospecting, 48
Goodyear, Charles, 15
Goodyear, Henry, 15, 172n2
Goodyear, Robert, 15, 172n2
Governor Marcy (steamer), 20
Gowen, John E., 46–49, 68, 78, 97,
    101, 113–114, 158–159
Grace A. Green (schooner), 129–
    130
Graham, William A., 40
Grant, Ephraim B., 78
Great Chain of the Hudson, 111
Green, Andrew, 61
Green, Ann, 64, 177n13
Green, John B., 59–66, 77–80, 83–90,
    92, 97–104, 119–120, 123–124,
    147
    alcoholism, 134–136
    autobiographies, 132–133, 135
    death, 161
    family background, 59–61
    marital problems, 134–135
Green, Lucy, 61, 147
Green, Mary, 61, 119
Green, Peter, 60–61

Griffith, John Morris, 49
guano industry, 113–114, 183n19
Gwynne pumps, 116

Hale, Alfred, 162
Hale, James W., 7, 9
Harnden's Express, 9, 76
Harrington, Elliot P., gallery 14,
    gallery 15, 91–92, 116–117,
    120–125, 143–145, 149–155
Hell Gate, New York City, 27, 34–35,
    76–78, gallery 5, 110, 157
Hendrick Hudson (steamer), 88
HMS Buzzard. See Buzzard, HMS
HMS Hussar. See Hussar, HMS
HMS Plumper. See Plumper, HMS
HMS Royal George. See Royal
    George, HMS
Holland, John Philip, 95
Howard & Ash, Co., 68, 177n3
Humboldt (steamer), 97
Hunley (submarine), 145, 158
Hussar, HMS, 27, 78, 114–115,
    156–157
Husted, Peter V., 110
Huzzar, HMS. See Hussar, HMS

Illinois (brig), 20
Illinois (canal scow), 62
Indian Queen (steamer), 20
International Submarine Mining
    Company, 166, 168
Irwin, William, 48

J. Wolcott (steamer), 49
Jaminet, Alphonse, 160
Jennings, Grace A., 100, 109,
    134–135, 161

Kendrick, William A., 130
Keystone State (steamer), 44
Kinney, T. E., 95
Knickerbocker (steamer), 40
Kroehl, Julius B., 110, 159–160

Lady (ship), 13
Lake Erie
    winter conditions, 18
    steamships, 176n1
lamp, underwater, 79, 107, 179n12
Lefferts, Rem, 43
Lexington (Long Island Sound
    steamer), 9, 48, 76, 124
Lexington (Lake Erie steamer), 86,
    89
Little Erie (steamer), 17–18
Lubin, Vivienne, 156

Macedonian, USS, 26, 173n9
Madison (steamer), 86, 89
Maillefert, Benjamin, 33–35, 76–81,
    gallery 3, gallery 14, 110, 146,
    157–158
Malakoff (schooner), 130
Marine Cigar (submarine), 39–40,
    gallery 9, 92–95, 110
marine compass. See mineral
    compass
Martin, Jacob, 150–151
Mary D. Leach (schooner), 166
Mayflower (steamer), 44
Maynard, John (fictional character),
    14
McCluer, Orrin, 18–20
McCracken, Samuel, 167
McDonnell, William, 85–86, gallery
    12, 91

McLaughlin, James H., 152–153,
185n6
Mexico (wreck), 8
Miles, George, Captain, 16
mineral compass, 18–19, 124, 166
Minnesota (steamer), 115
Mississippi River Squadron, 147
Missouri, USS, 29–30, 44–46, 48–49,
175n11
Morse, Andrew J., 162
Mosquito Fleet, 29–30

Nautilus diving bells, 36, 75, 80,
gallery 7–8, 111–113, 159
Nautilus (Fulton submarine), 37
Nelson, William, 41, 84, 141
New York Aquarium, 156
New York Sub-Marine Armour
Company, 7–8
Newton, William, 120, 153
Newell, Seth P., 49
Niblo's Garden, New York City, 7,
171n9
Northerner (steamer), 117

Ocean (steamer), 94
Ogdensburg (propeller), 72–73,
gallery 12, 93
Oneida (propeller), 92
Oregon (propeller), 116
Osborn, Miles, 121

palace steamers, 57–58, 141
Pancoast, Seth, 168
panoramas, 131–132
Pasley, Charles William, General Sir,
8, 33
Patchin, Aaron D., 43

Payne, William, 152
pearl harvesting, 5–6, 171n4–5
Peck, Frederick M., 145
Pensacola, Florida, 96, 115
Petty, J. B., Captain, 72
Philadelphia Submarine Mining
Company, 159
Phillips, James H., gallery 15, 116,
153
Phillips, Lodner D., 38–40, gallery
8–9, 92–95, 110–111, 145, 158
Pike, Charles F., 166–168
Pioneer (steamer), 88
piracy, 3–5
Ploughboy (steamer), 109
Plumper, HMS, 47
Powers, Hiram, 46
Pratt, Charles B., 31–32, 77–78,
gallery 16, 114–115, 156–157,
162–163
Pratt, Lucius H., 43
Prince Alfred (gunboat), 151
Princeton (propeller), 115–116
privateering, 3–5
propeller design, 26, 58

Quigley, Martin ("Perglich"), 69–71,
77, 84–86, 89–91, 115–116, 120,
142–143, 156

Reed, Charles Manning, 11–12, 14,
43
Robinson, Edward R., 30–32, 44–46
Roby, Charles C., 50–51
Rocky Mountains (ship), 15
Royal George, HMS, 33
Rush (brig), 111
Ryerson, Van Buren, 159

San Pedro de Alcantara (ship), 78, 96, 101–102, 112
Sandusky (barque), 20
Scott, William, Major, 31
Schrader, August, 162
Sears, Henry Beaufort, 35–36, 75, 80, 111–112, 159
Sears, Richard, 49
Sebastopol, Crimea, 158–159
Siebe, Augustus, 6, 48, gallery 16, 175n10
Silver Bank reef, 101–103, 135
Silver Lake Monster, 105–106
slave trade, 28, 42
Smith, Andrew H., 160
Southerner (steamer), 89
Spitfire (ship), 28–29
St. Louis, Missouri, 156
    Eads' Bridge, 160
    Fire of 1849, 42
    Great Ice Jam of 1856, 117
Staffordshire (ship), 97
Star (steamer), 16
Stebbins, Daniel R., 49–53, 67–68, 83–84, 177n2
Stevens, Charles, 129
Sub Marine Explorer (submarine), 159
Sub Marine Explorer (diving bell), 159
submarine armor, 6–9, 15–16, 44, 47–48, 67–68, 78, gallery 2, gallery 10, gallery 11, gallery 16, 101, 161–162, 178n7
Submarine No. 1 (steamer), 41
Submarine No. 2 (steamer), 41
Submarine No. 3 (steamer), 41
Submarine No. 4 (steamer), 42, 116–117
Submarine No. 7 (steamer), 146

Submarine No. 12 (steamer), 142
submarine vessels, 36–40, gallery 8, 145–146, 163
Sultana (steamer), 130

Taylor, George W., 8–9, 15–18, 27–32, 41, 172n12
Taylor, William Hannis, 15, 26–27, gallery 2–3, 161
    as privateer, 3–5, 171n1
    coopering, 23–24
    electric motor, 8–9, 23
    salvage of *Telemaque*, 24–26
    submarine armor, 6–9
Telemaque (barge), 24–26, gallery 3, 173n3
Titus, Thomas (Captain), 12–14
Toledo (brig), 20
Tope, John, gallery 12, 98–99, 181n6
torpedoes, 27–28, gallery 14, 139, 146
Tremble, Theophilus Peter, 119, 147, 183n1
Turner, Daniel (Captain), 3–5, 11
Turtle (submarine), 37

USS *Delaware*. See *Delaware*, USS
USS *Erie*. See *Erie*, USS
USS *Macedonian*. See *Macedonian*, USS
USS *Missouri*. See *Missouri*, USS

Vandalia (propeller), 58
Vermillion (freighter), 168–169
Villeroi, Brutus. *See* De Villeroi, Brutus
Virginia, CSS, 154

Wallace (barque), 102
Ward, Eber, 68, 71, 75, 78, 135

Ward, Samuel, 68, 71, 75

Watt, Francis M. *See* Lubin, Vivienne

Welles, Gideon, 145

Wells, Fargo and Company, 76

Wells, Henry, 75-77, 80-81, 88-90, 93, 125

Wells, Thomas F., 47-48, 78, gallery 10, gallery 11, 97, 99, 101, 114

Western River Improvement and Wrecking Company, 142

Whipple, James A., 30-32, 44-46, 78, 88-90, gallery 4, 95-96, 101-102, 157, 175n10

White Cloud (steamer), 42

Willard Johnson (schooner), 149

Williams, William, Captain, 59, 63

Wisconsin (steamer), 15

Wolcott, James, 49

wreckmasters, 65

Yorktown (schooner), 107

Young America (steamer), 116

Young Lyon (schooner), 20

Zalinski, Edmund L., 94-95